THE SOLAR SYSTEM

MARS

REVISED EDITION

Linda T. Elkins-Tanton

Facts On File
An imprint of Infobase Publishing

To Marc Parmentier and his research group at Brown University,
for their exceptional gifts as scientists and colleagues,
and in recognition of the exciting and rewarding research
about Mars that we have done together

MARS, Revised Edition

Copyright © 2011, 2006 by Linda T. Elkins-Tanton

Facts On File, Inc.
An imprint of Infobase Publishing
132 West 31st Street
New York NY 10001

Library of Congress Cataloging-in-Publication Data
Elkins-Tanton, Linda T.
    Mars / author, Linda T. Elkins-Tanton; foreword, Maria T. Zuber.—Rev. ed.
        p. cm.
    Includes bibliographical references and index.
    ISBN 978-0-8160-7699-4 (alk. paper)
    1. Mars (Planet)—Popular works. I. Title.
QB641.E43 2010
523.43—dc22   2009050363

Facts On File books are available at special discounts when purchased in bulk quantities for businesses, associations, institutions, or sales promotions. Please call our Special Sales Department in New York at (212) 967-8800 or (800) 322-8755.

You can find Facts On File on the World Wide Web at http://www.factsonfile.com

Text design by Annie O'Donnell
Composition by Hermitage Publishing Services
Illustrations by Richard Garratt
Photo research by Elizabeth H. Oakes
Cover printed by Bang Printing, Brainard, Minn.
Book printed and bound by Bang Printing, Brainard, Minn.
Date printed: November 2010
Printed in the United States of America

10 9 8 7 6 5 4 3 2 1

This book is printed on acid-free paper.

Contents

Appendix 3: A List of All Known Moons   242

Foreword

While I was growing up, I got my thrills from simple things—one was the beauty of nature. I spent hours looking at mountains, the sky, lakes, et cetera, and always seeing something different. Another pleasure came from figuring out how things work and why things *are* the way they *are*. I remember constantly looking up things from why airplanes fly to why it rains to why there are seasons. Finally was the thrill of discovery. The excitement of finding or learning about something new—like when I found the Andromeda galaxy for the first time in a telescope—was a feeling that could not be beat.

Linda Elkins-Tanton's multivolume set of books about the solar system captures all of these attributes. Far beyond a laundry list of facts about the planets, the Solar System is a set that provides elegant descriptions of natural objects that celebrate their beauty, explains with extraordinary clarity the diverse processes that shaped them, and deftly conveys the thrill of space exploration. Most people, at one time or another, have come across astronomical images and marveled at complex and remarkable features that seemingly defy explanation. But as the philosopher Aristotle recognized, "Nature does nothing uselessly," and each discovery represents an opportunity to expand human understanding of natural worlds. To great effect, these books often read like a detective story, in which the 4.5-billion year history of the solar system is reconstructed by integrating simple concepts of chemistry, physics, geology, meteorology, oceanography, and even biology with computer simulations, laboratory analyses, and the data from the myriad of space missions.

Starting at the beginning, you will learn why it is pretty well understood that the solar system started as a vast, tenuous ball of gas and dust that flattened to a disk with most of the mass—the future Sun—at the center. Much less certain is the transition from a dusty disk to the configuration with the planets, moons, asteroids, and comets that we see today. An ironic contrast is the extraordinary detail in which we understand some phenomena, like how rapidly the planets formed, and how depressingly uncertain we are about others, like how bright the early Sun was.

Once the planets were in place, the story diverges into a multitude of fascinating subplots. The oldest planetary surfaces preserve the record of their violent bombardment history. Once dismissed as improbable events, we now know that the importance of planetary impacts cannot be overstated. One of the largest of these collisions, by a Mars-sized body into the Earth, was probably responsible for the formation of the Earth's Moon, and others may have contributed to extinction of species on Earth. The author masterfully explains in unifying context the many other planetary processes, such as volcanism, faulting, the release of water and other volatile elements from the interiors of the planets to form atmospheres and oceans, and the mixing of gases in the giant planets to drive their dynamic cloud patterns.

Of equal interest is the process of discovery that brought our understanding of the solar system to where it is today. While robotic explorers justifiably make headlines, much of our current knowledge has come from individuals who spent seemingly endless hours in the cold and dark observing the night skies or in labs performing painstakingly careful analyses on miniscule grains from space. Here, these stories of perseverance and skill receive the attention they so richly deserve.

Some of the most enjoyable aspects of these books are the numerous occasions in which simple but confounding questions are explained in such a straightforward manner that you literally feel like you knew it all along. How do you know what is inside a planetary body if you cannot see there? What makes solar system objects spherical as opposed to irregular in shape? What causes the complex, changing patterns at the top of Jupiter's atmosphere? How do we know what Saturn's rings are made of?

When it comes right down to it, all of us are inherently explorers. The urge to understand our place on Earth and the extraordinary worlds beyond is an attribute that makes us uniquely human. The discoveries so lucidly explained in these volumes are perhaps most remarkable in the sense that they represent only the tip of the iceberg of what yet remains to be discovered.

—Maria T. Zuber, Ph.D.
E. A. Griswold Professor of Geophysics
Head of the Department of Earth,
Atmospheric and Planetary Sciences
Massachusetts Institute of Technology
Cambridge, Massachusetts

Preface

On August 24, 2006, the International Astronomical Union (IAU) changed the face of the solar system by dictating that Pluto is no longer a planet. Though this announcement raised a small uproar in the public, it heralded a new era of how scientists perceive the universe. Our understanding of the solar system has changed so fundamentally that the original definition of *planet* requires profound revisions.

While it seems logical to determine the ranking of celestial bodies by size (planets largest, then moons, and finally asteroids), in reality that has little to do with the process. For example, Saturn's moon Titan is larger than the planet Mercury, and Charon, Pluto's moon, is almost as big as Pluto itself. Instead, scientists have created specific criteria to determine how an object is classed. However, as telescopes increase their range and computers process images with greater clarity, new information continually challenges the current understanding of the solar system.

As more distant bodies are discovered, better theories for their quantity and mass, their origins, and their relation to the rest of the solar system have been propounded. In 2005, a body bigger than Pluto was found and precipitated the argument: Was it the 10th planet or was, in fact, Pluto not even a planet itself? Because we have come to know that Pluto and its moon, Charon, orbit in a vast cloud of like objects, calling it a planet no longer made sense. And so, a new class of objects was born: the dwarf planets.

Every day, new data streams back to Earth from satellites and space missions. Early in 2004, scientists proved that standing liquid water once existed on Mars, just a month after a mission visited a comet and discovered that the material in its nucleus is as strong as some *rocks* and not the loose pile of ice and dust expected. The MESSENGER mission to Mercury, launched in 2004, has thus far completed three flybys and will enter Mercury orbit at 2011. The mission has already proven that Mercury's core is still molten, raising fundamental questions about processes of planetary evolution, and it has sent back to Earth intriguing information about the composition of Mercury's crust. Now the New Horizons mission is on its way to make the first visit to Pluto and the Kuiper belt. Information arrives from space observations and Earth-based experiments, and scientists attempt to explain what they see, producing a stream of new hypotheses about the formation and evolution of the solar system and all its parts.

The graph below shows the number of moons each planet has; large planets have more than small planets, and every year scientists discover new bodies orbiting the gas giant planets. Many bodies of substantial size orbit in the asteroid belt, or the Kuiper belt, and many sizable asteroids cross the orbits of planets as they make their way around the Sun. Some planets' moons are unstable and will in the near future (geologically speaking) make new ring systems as they crash into their hosts. Many moons, like Neptune's giant Triton, orbit their planets backward (clockwise when viewed from the North Pole, the opposite way that the planets orbit the

The mass of the planet appears to control the number of moons it has; the large outer planets have more moons than the smaller, inner planets.

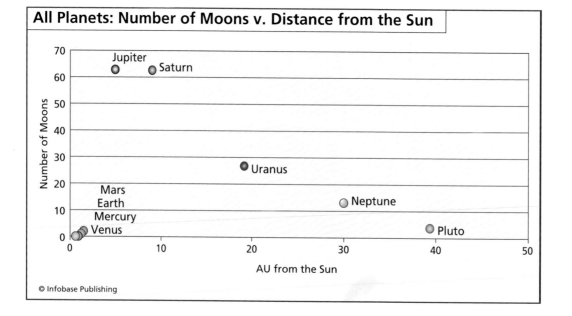

All Planets: Number of Moons v. Distance from the Sun

© Infobase Publishing

Sun). Triton also has the coldest surface temperature of any moon or planet, including Pluto, which is much farther from the Sun. The solar system is made of bodies in a continuum of sizes and ages, and every rule of thumb has an exception.

Perhaps more important, the solar system is not a static place. It continues to evolve—note the drastic climate changes we are experiencing on Earth as just one example—and our ability to observe it continues to evolve, as well. Just five planets visible to the naked eye were known to ancient peoples: Mercury, Venus, Mars, Jupiter, and Saturn. The Romans gave these planets the names they are still known by today. Mercury was named after their god Mercury, the fleet-footed messenger of the gods, because the planet Mercury seems especially swift when viewed from Earth. Venus was named for the beautiful goddess Venus, brighter than anything in the sky except the Sun and Moon. The planet Mars appears red even from Earth and so was named after Mars, the god of war. Jupiter is named for the king of the gods, the biggest and most powerful of all, and Saturn was named for Jupiter's father. The ancient Chinese and the ancient Jews recognized the planets as well, and the Maya (250–900 C.E., Mexico and environs) and Aztec (~1100–1700 C.E., Mexico and environs) knew Venus by the name Quetzalcoatl, after their god of good and light, who eventually also became their god of war.

Science is often driven forward by the development of new technology, allowing researchers to make measurements that were previously impossible. The dawn of the new age in astronomy and study of the solar system occurred in 1608, when Hans Lippenshey, a Dutch eyeglass-maker, attached a lens to each end of a hollow tube and thus created the first telescope. Galileo Galilei, born in Pisa, Italy, in 1564, made his first telescope in 1609 from Lippenshey's model. Galileo soon discovered that Venus has phases like the Moon does and that Saturn appeared to have "handles." These were the edges of Saturn's rings, though the telescope was not strong enough to resolve the rings correctly. In 1610, Galileo discovered four of Jupiter's moons, which are still called the Galilean satellites. These four moons were the proof that not every heavenly body orbited the Earth as Ptolemy, a Greek philosopher, had asserted around 140 C.E. Galileo's discovery

Obliquity, Inclination, Rotation

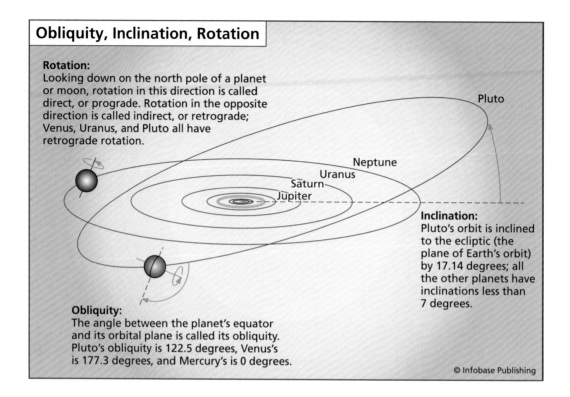

Rotation:
Looking down on the north pole of a planet or moon, rotation in this direction is called direct, or prograde. Rotation in the opposite direction is called indirect, or retrograde; Venus, Uranus, and Pluto all have retrograde rotation.

Pluto

Neptune
Uranus
Saturn
Jupiter

Inclination:
Pluto's orbit is inclined to the ecliptic (the plane of Earth's orbit) by 17.14 degrees; all the other planets have inclinations less than 7 degrees.

Obliquity:
The angle between the planet's equator and its orbital plane is called its obliquity. Pluto's obliquity is 122.5 degrees, Venus's is 177.3 degrees, and Mercury's is 0 degrees.

© Infobase Publishing

Obliquity, orbital inclination, and rotational direction are three physical measurements used to describe a rotating, orbiting body.

was the beginning of the end of the strongly held belief that the Earth is the center of the solar system, as well as a beautiful example of a case where improved technology drove science forward.

The concept of the Earth-centered solar system is long gone, as is the notion that the heavenly spheres are unchanging and perfect. Looking down on the solar system from above the Sun's north pole, the planets orbiting the Sun can be seen to be orbiting counterclockwise, in the manner of the original *protoplanetary disk* of material from which they formed. (This is called *prograde* rotation.) This simple statement, though, is almost the end of generalities about the solar system. Some planets and dwarf planets spin backward compared to the Earth, other planets are tipped over, and others orbit outside the *ecliptic* plane by substantial angles, Pluto in particular (see the following figure on *obliquity* and orbital *inclination*). Some planets and moons are still hot enough to be volcanic, and some produce *silicate* lava (for example, the Earth and Jupiter's moon Io), while others have exotic lavas made of molten ices (for example, Neptune's moon Triton).

Today, we look outside our solar system and find planets orbiting other stars, more than 400 to date. Now our search for signs of life goes beyond Mars and Enceladus and Titan and reaches to other star systems. Most of the science presented in this set comes from the startlingly rapid developments of the last 100 years, brought about by technological development.

The rapid advances of planetary and heliospheric science and the astonishing plethora of images sent back by missions motivate the revised editions of the Solar System set. The multivolume set explores the vast and enigmatic Sun at the center of the solar system and moves out through the planets, dwarf planets, and minor bodies of the solar system, examining each and comparing them from the point of view of a planetary scientist. Space missions that produced critical data for the understanding of solar system bodies are introduced in each volume, and their data and images shown and discussed. The revised editions of *The Sun, Mercury, and Venus, The Earth and the Moon,* and *Mars* place emphasis on the areas of unknowns and the results of new space missions. The important fact that the solar system consists of a continuum of sizes and types of bodies is stressed in the revised edition of *Asteroids, Meteorites, and Comets.* This book discusses the roles of these small bodies as recorders of the formation of the solar system, as well as their threat as *impactors* of planets. In the revised edition of *Jupiter and Saturn,* the two largest planets are described and compared. In the revised edition of *Uranus, Neptune, Pluto, and the Outer Solar System,* Pluto is presented in its rightful, though complex, place as the second-largest known of a extensive population of icy bodies that reach far out toward the closest stars, in effect linking the solar system to the Galaxy itself.

This set hopes to change the familiar and archaic litany *Mercury, Venus, Earth, Mars, Jupiter, Saturn, Uranus, Neptune, Pluto* into a thorough understanding of the many sizes and types of bodies that orbit the Sun. Even a cursory study of each planet shows its uniqueness along with the great areas of knowledge that are unknown. These titles seek to make the familiar strange again.

Acknowledgments

Foremost, profound thanks to the following organizations for the great science and adventure they provide for humankind and, on a more prosaic note, for allowing the use of their images for these books: the National Aeronautics and Space Administration (NASA) and the National Oceanic and Atmospheric Administration (NOAA), in conjunction with the Jet Propulsion Laboratory (JPL) and Malin Space Science Systems (MSSS). A large number of missions and their teams have provided invaluable data and images, including the *Solar and Heliospheric Observer (SOHO), Mars Global Surveyor (MGS), Mars Odyssey,* the *Mars Exploration Rovers (MERs), Galileo, Stardust, Near-Earth Asteroid Rendezvous (NEAR),* and *Cassini.* Special thanks to Steele Hill, SOHO Media Specialist at NASA, who prepared a number of images from the SOHO mission, to the astronauts who took the photos found at Astronaut Photography of the Earth, and to the providers of the National Space Science Data Center, Great Images in NASA, and the NASA/JPL Planetary Photojournal, all available on the Web (addresses given in the reference section).

Many thanks also to Frank K. Darmstadt, executive editor; to Elizabeth Oakes, photo researcher; to Jodie Rhodes, literary agent; and to E. Marc Parmentier at Brown University for his generous support.

Introduction

The international space community has reached the point where it is able to monitor Mars around the clock; Mars and the Moon are covered continuously from multiple points of view. The National Aeronautics and Space Administration (NASA) had three orbiters at work in 2009: *2001 Mars Odyssey, Mars Global Surveyor,* and *Mars Reconnaissance Orbiter.* The European Space Agency (ESA) has its orbiter *Mars Express* functioning well, and on the surface the extremely long-lived Mars rovers *Spirit* and *Opportunity* continue to investigate the fine details of the surface. In the last year, the Mars *Phoenix* lander added a wealth of information about water in the soils of the planet and the chemical past of the surface, and NASA has planned a highly ambitious surface rover called *Mars Science Laboratory* for launch in 2011.

Mars exploration has never been more active, and our understanding of the planet is advancing rapidly. The primary questions remain. Is there life on Mars now? Did life exist on Mars in the past? And if not, why not? Dry river valleys are spread across wide regions of the planet; they have been recognized for years. Now, new discoveries reveal gullies carved by recent groundwater flow, thick ice deposits protected by rocks and soil even at the equator, and new evidence for lakes and seas in Mars's past. Mars's surface has highly acidic regions, probably created by interactions between water and volcanic emissions, and it has other regions dominated by carbonate minerals, which are formed by interaction with water and are destroyed by acids. Mars apparently has a highly variable surface, and everywhere its surface is the product of interactions with water.

Together this creates the picture of a planet with a wet, hospitable past, rapid climate change, and a dry, cold present.

If Mars lost its water and became a desert planet, could Earth follow the same path? And if there was once life on Mars, did the climate change cause its extinction? In addition to these compelling questions, the Martian surface has some of the oldest planetary crust in the solar system, containing clues to conditions in early planets that cannot be obtained elsewhere. Finally, Mars is a possible destination for human travel, the next attainable target after the Moon.

Sending missions to Mars has been an exceptionally risky venture. Between 1960 and 1995, the United States and the Soviet Union (since 1991, Russia) launched a total of 27 missions to Mars, 15 of which failed completely, three of which returned some data but did not complete their missions, and nine of which were successes. (All these missions are listed in a table in chapter 7.) The Soviet Union suffered a particularly demoralizing string of 10 consecutive failures. A one-third chance of overall success is not enticing when each mission costs hundreds of millions of dollars or more. Until the 1990s, most of what mankind knew of Mars came from the two *Viking* landers that the United States sent to Mars in 1975. This photo below of the rocky surface of Mars and the foot of the *Viking 1* lander was the first image ever taken from the surface of Mars on July 20, 1976.

Mars, Revised Edition begins with a discussion of Mars as a planet in orbit as can be seen in the figure on page xvii), covering fundamental facts such as its mass and size and discussing its seasons with their surface effects on the planet and the

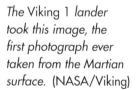

The Viking 1 *lander took this image, the first photograph ever taken from the Martian surface.* (NASA/Viking)

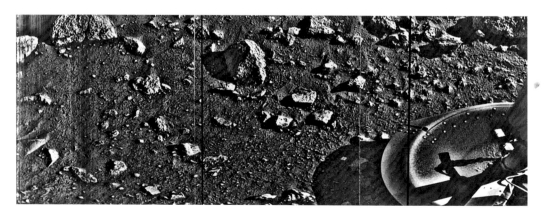

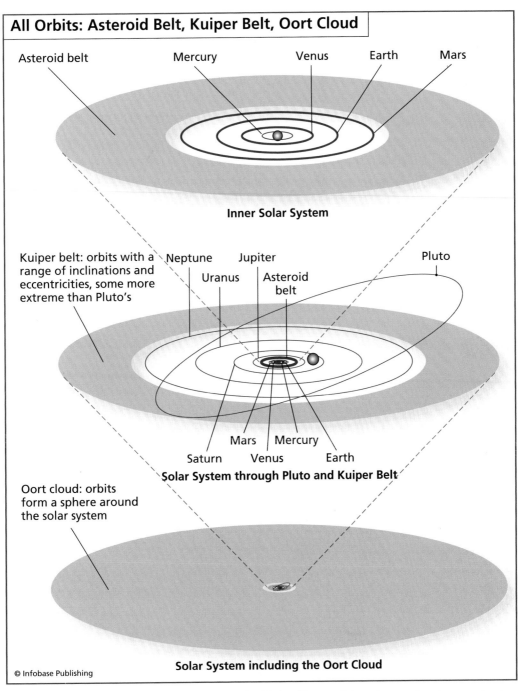

All Orbits: Asteroid Belt, Kuiper Belt, Oort Cloud

Asteroid belt Mercury Venus Earth Mars

Inner Solar System

Kuiper belt: orbits with a range of inclinations and eccentricities, some more extreme than Pluto's

Neptune Jupiter Pluto

Uranus Asteroid belt

Mars Mercury

Saturn Venus Earth

Solar System through Pluto and Kuiper Belt

Oort cloud: orbits form a sphere around the solar system

© Infobase Publishing

Solar System including the Oort Cloud

This book covers the planet Mars, whose orbit is highlighted here. All the orbits are far closer to circular than shown in this oblique view, which was chosen to show the inclination of Pluto's orbit to the ecliptic.

theories on how the severity of its seasons have changed over time. The book continues by discussing what is known about the material that makes up Mars. The formation and early evolution of the Moon and the Earth have both been inferred as far as scientists have been able by analyzing the compositions of rocks, but for Mars the available samples are far fewer.

A very few of the many meteorites that have fallen to Earth have been proven, amazingly, to have come from Mars. As of 2009 there are 49 known meteorites from Mars. These are the only direct samples from the planet since no space missions have returned to Earth with material. From these few samples and the information sent back to Earth from Mars missions, scientists have been able to make comparisons between the compositions of Mars and the Earth and to begin to piece together the geological history of the planet. Chapter 2 covers what is known and inferred about the composition and structure of the interior of Mars, including theories of planetary formation, of an unusually iron-rich *mantle,* and of the planet's paradoxically strong early magnetic field, which has now completely ended.

Chapter 3 includes discussions of weather, surface conditions, and surface features. Though Venus is Earth's closest neighbor in terms of distance from the Sun, size, and density, Mars is more nearly the twin of Earth in other respects. Its dusty surface covered with rocks and sand dunes resembles terrestrial deserts and carries the familiar marks of past rivers, glaciers, and even oceans. The panoramic image of the Martian surface in the photo on page xx was taken by the *Spirit* rover on April 3, 2004, although one could imagine it being in any desert here on Earth. Mars's atmosphere is thin and consists of carbon dioxide and nitrogen rather than Venus's thick sulfuric acid-laden blanket.

Despite the dusty desert image, closer attention shows that the composition of the surface is highly variable, from *igneous rocks* from volcanoes to clay minerals caused by chemical reaction with water in the past. The combination of landers and orbiters shows the complexity of surface conditions with increasing clarity. Mars missions, in particular the Mars Global Surveyor, Mars Express, and Mars Reconnaissance

Orbiter, have created intricately detailed topographic and compositional maps of Mars and taken tens of thousands of images of the surface. Photogeologists studying the images have been able to show evidence for flowing water in the distant past and groundwater seepage even in the present, opening the possibility that the surface of Mars was hospitable to life at one time, the topic of chapter 4.

In mankind's never-ending search for life on other planets, Mars has always ranked among the most likely places to look, initially because of its nearness to the Earth and easily visible surface. In the 19th century, astronomers searched for life on Mars simply because they could: Other than the Moon, it was the only solar system body that could be examined closely through the telescopes of the day. Features suggesting life were immediately found in the shape of greenish patches resembling vegetation. Soon after that, a mistranslation from Italian on a map of Mars compelled the American astronomer Percival Lowell to begin a decades-long investigation of linear features that he claimed were canals built by intelligent civilizations. Lowell's investigations set off a near-global frenzy about little green men from Mars, the vestiges of which are still with us today in the form of the novels of Ray Bradbury and others,

This enhanced color view of the eastern rim and floor of Victoria Crater in the Meridiani Planum region comes from the High Resolution Imaging Science Experiment (HiRISE) camera on Mars Reconnaissance Orbiter: These ridges may be fractures surrounded by chemically cemented sedimentary bedrock, implying water activity in the past. (NASA/JPL/University of Arizona)

The Mars Exploration Rover Spirit *took this panoramic image of the desolate, freezing surface of Mars before it rolled off the lander.* (NASA/ JPL/Cornell University)

the cartoon character Marvin the Martian, and a myriad of other examples.

Though Mars does not have canals or any other evidence for intelligent life, it was in the past a promising location for the development of life. New and existing Mars missions may provide definitive evidence as to whether or not there was once life on Mars. In 1996, a new era of Mars research began with the successful NASA missions Mars Global Surveyor and Mars Pathfinder. Between 1996 and 2007, the United States, Russia, Japan, and the ESA launched a total of 13 missions to Mars. These fared better than their predecessors: seven were successes, five were failures, and one was a partial success (the Mars Express mission, which lost its *Beagle* lander but retained its orbiter). In 1998 and 1999, the United States suffered another series of painful losses as the Mars Climate Orbiter, Mars Polar Lander, and Deep Space 2 missions all failed. Despite the success of the *Mars Odyssey* orbiter, the future funding for Mars missions from the United States was in serious doubt. There was even discussion in the planetary community that the credibility of the space program itself was in question.

With the startling successes of the Mars Exploration Rover *A (Spirit)* and *B (Opportunity)*, *Mars Reconnaissance Orbiter*, and the *Phoenix* lander, support, confidence, and interest in the American space program and especially in Mars exploration have surged. These two rovers bounced along the Martian surface on giant landing balloons, unfolded themselves, and rolled away into the Martian landscape with hardly a hitch and continue to operate years after their expected lifetimes. The recovery of the Mars program and its new structure are described in chapters 6 and 7.

The quantities of new data and attendant new theories of Martian conditions and evolution make writing a book about Mars a special challenge. New ideas are being published every day. The similarities between the Earth and Mars make Mars particularly interesting, and the flood of attention and data certainly qualifies Mars to command its own book in this Solar System set.

Mars: Fast Facts about a Planet in Orbit

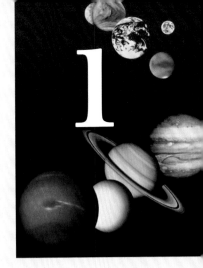

Planetary sciences can be divided roughly into research topics that concern physics and research topics that concern chemistry. In geoscience departments all over the world, scientists are also divided this way, defined as geophysicists or geochemists and petrologists (petrology is the study of minerals and rocks). In truth most research topics require some understanding of both disciplines. Studying planetary formation, for example, requires a knowledge of the physics of impacts, gravity, and heating and also a knowledge of chemical compounds, minerals, and reactions. The topics of orbits and rotations, though, lie almost purely in the field of physics. The shape of a planetary body depends upon the speed of its rotation, the number and size of other bodies nearby, and the density and strength of the material that composes it. Shapes and speeds can be measured readily and used to help determine characteristics of the composition. The behavior of an orbit is likewise the product of the forces on the planetary body: The orbit is shaped and its speed determined by the masses of the orbiting body and the Sun and by gravitational pulls from nearby bodies. In turn, the shape and size of the orbit along with the tilt of the planet's axis help determine the range of temperatures experienced by the planet—in particular, its seasons. Details of Mars's orbit are given in the table on page 3.

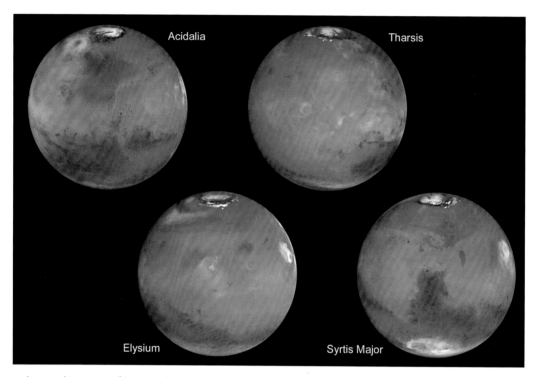

Taking advantage of Mars's closest approach to Earth in eight years, astronomers used NASA's Hubble Space Telescope to take these images in 1997. Each view depicts the planet as it completes one-quarter of its daily rotation. The north polar cap is tilted toward the Earth and is visible at the top of each picture, shrunk to its smallest size by high northern summer. The upper-left image is centered near the location of the Pathfinder landing site. Dark sand dunes surround the polar cap in the Acidalia region. Below and to the left of Acidalia are the massive canyon systems of Valles Marineris. A large cyclonic storm composed of water ice is churning near the polar cap. The upper-right image is centered on Tharsis. The bright, ringlike feature just to the left of center is the volcano Olympus Mons, more than 340 miles (550 km) across and 17 miles (27 km) high. Thick deposits of fine-grained, windblown dust cover most of this hemisphere. Prominent late afternoon clouds hang along the planet's right limb. The lower-left image is centered near the Elysium volcanic region. In the upper left of this image, at high northern latitudes, a large chevron-shaped area of water ice clouds marks a storm front. Along the right limb, a large cloud system has formed around the Olympus Mons volcano. The lower-right image is centered on the low, dark shield volcano known as Syrtis Major, first seen telescopically by the astronomer Christiaan Huygens in the 17th century. To the south of Syrtis is the large circular Hellas impact crater. Along the right limb, late afternoon clouds have formed around the volcano Elysium. (NASA/Malin Science Systems)

MARS'S ORBIT	
rotation on its axis ("day")	1.02596 Earth days
rotation direction	prograde (counterclockwise when viewed from above the North Pole)
ellipticity	0.005, meaning the planet's equator is about one half percent longer than its polar radii
sidereal period ("year")	1.88 Earth years
orbital velocity (average)	14.99 miles/second (24.13 km/sec)
sunlight travel time (average)	12 minutes and 40 seconds to reach Mars
average distance from the Sun	141,637,134 miles (227,936,640 km), or 1.52 AU
perihelion	128,378,798 miles (206,600,000 km), or 1.381 AU from the Sun
aphelion	154,849,935 miles (249,200,000 km), or 1.666 AU from the Sun
orbital eccentricity	0.0934 (unitless; see below)
orbital inclination to the ecliptic	1.85 degrees
obliquity (inclination of equator to orbit)	25.2 degrees, similar to Earth's, and therefore Mars now has seasons somewhat like the Earth's

The sidebar "Fundamental Information about Mars," as well as the tables of physical and orbital data in this chapter, contains information largely calculated by scientists who think about physics. Based on these data, Mars and the Earth seem similar: They are at similar distances from the Sun, have days about the same length, years that differ by less than a factor of two in length, similar density, and similar obliquity (inclination of the planet's equator to its orbit). This list of similarities would make someone new to planetary science think that the planets should be similar in other ways, but of course Mars is freezing cold, waterless, and covered with dust storms, while the Earth is temperate, covered with liquid water, and has a variety of weathers. How did Mars evolve to be so different from the Earth, when their basic physical parameters are so similar?

First, to stay in the realm of physics, Mars has two small moons while the Earth has one large one. Earth's Moon stabilizes its rotation for long periods of time, preventing Earth's rotation axis from tipping farther over or becoming much

FUNDAMENTAL INFORMATION ABOUT MARS

The Moon, Mars, and the Earth fall into a regular size order: Mars's radius is almost exactly twice that of the Moon, and Earth's is almost exactly twice that of Mars. Since the volume of a sphere is related to the cube of its radius

$$V = \frac{4}{3}\pi r^3,$$

the bodies are far more different in terms of volume: Mars has 0.13 times the Earth's volume, and the Moon again 0.13 times Mars's volume. Though in many ways Mars

FUNDAMENTAL FACTS ABOUT MARS	
equatorial radius	2,110.36 miles (3,396.200 km), or 0.53 times Earth's
north polar radius	2,097.92 miles (3,376.19 km)
south polar radius	2,101.90 miles (3,382.58 km) (these precise numbers are thanks to the Mars Orbital Laser Altimeter)
volume	3.91×10^{10} cubic miles (1.63×10^{11} km^3)
mass	1.41×10^{24} lbs (6.42×10^{23} kg)
average density	246 lb/ft^3 (3,940 kg/m^3)
acceleration of gravity on the surface at the equator	12.2 ft/sec^2 (3.69 m/sec^2), or 0.38 times Earth's
magnetic field strength at the surface	patchy but strong in places, varying from 10^{-4} tesla (stronger than the Earth's) to 10^{-9} tesla
rings	0
moons	2

more upright. Mars, however, has no such stabilizing force. Over time Mars is thought to have tipped back and forth, creating far more extreme seasons than it now has. Liquid water and an atmosphere are necessary for a climate like Earth's,

resembles the Earth on its surface (deserts, volcanic flows, dry river valleys, ice caps), the small volume and resulting small internal pressures of Mars make its internal processes considerably different from the Earth's, as can be seen on page 7.

Mars's average internal density is only about 71 percent of the Earth's, since its smaller mass does not press its internal matter into forms as dense as they are in the deep Earth. Similarly, Mars's gravity field is only about 40 percent as strong as the Earth's. These and other physical parameters for Mars are listed in the table at left.

Each planet and some other bodies in the solar system (the Sun and certain asteroids) have been given its own symbol as a shorthand in scientific writing. The symbol for Mars is shown below.

Symbol for Mars

© Infobase Publishing

Many solar system objects have simple symbols; this is the symbol for Mars.

though, and Mars has little or none of either. How Mars lost its water (or, more likely, had all its water frozen deeply into its *crust*) and its atmosphere are the subjects of considerable study, since understanding their loss may help understand how to keep the Earth's climate stable.

Climate is affected by many aspects of a planet, including its orbit and shape. A planet's rotation prevents it from being a perfect sphere. Spinning around an axis creates forces that cause the planet to swell at the equator and flatten slightly at the poles. Planets are thus shapes called oblate spheroids, meaning that they have different equatorial radii and polar radii, as shown in the image here. If the planet's equatorial radius is called r_e, and its polar radius is called r_p, then its flattening (more commonly called ellipticity, e) is defined as

$$e = \frac{r_e - r_p}{r_e}.$$

Ellipticity is the measure of by how much a planet's shape deviates from a sphere.

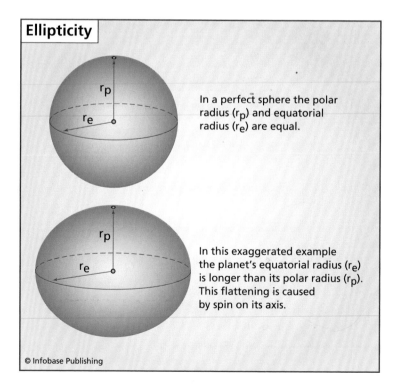

Ellipticity

In a perfect sphere the polar radius (r_p) and equatorial radius (r_e) are equal.

In this exaggerated example the planet's equatorial radius (r_e) is longer than its polar radius (r_p). This flattening is caused by spin on its axis.

© Infobase Publishing

The larger radius, the equatorial, is also called the *semimajor axis,* and the polar radius is called the *semiminor axis.* The Earth's semimajor axis is 3,960.8 miles (6,378.14 km), and its semimajor axis is 3,947.5 miles (6,356.75 km), so its ellipticity is (see figure on page 6)

$$e = \frac{3,963.19 - 3,950.01}{3,963.19} = 0.00333.$$

Because every planet's equatorial radius is longer than its polar radius, the surface of the planet at its equator is farther from the planet's center than the surface of the planet at the poles. To a lesser extent, the distance from the surface to the center of the planet changes according to topography such as mountains or valleys. Being at a different distance from the center of the planet means there is a different amount of mass between the surface and the center of the planet. What effect does the mass have? Mass pulls with its gravity. (For more information on gravity, see the sidebar "What Makes Gravity?" on page 84.) At the equator, where the radius of the

In this composite image, Earth and Mars can be compared directly: The size and water content of the Earth are apparent. Earth's image was acquired from the Galileo orbiter at about 6:10 A.M. PST on December 11, 1990, when the spacecraft was about 1.3 million miles (ca. 2.1 million km) from the planet during the first of two Earth flybys on its way to Jupiter. The Mars Global Surveyor acquired Mars's image in April 1999. (NASA/JPL)

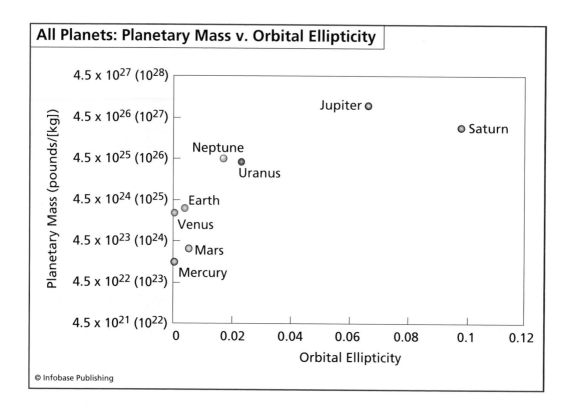

All Planets: Planetary Mass v. Orbital Ellipticity

© Infobase Publishing

The ellipticities of the planets differ largely as a function of their composition's ability to flow in response to rotational forces.

planet is larger, and therefore the amount of mass between the surface and the center of the planet is relatively larger, the pull of gravity is actually stronger than it is at the poles. Gravity is not a perfect constant on any planet: Variations in radius, topography, and the density of the material underneath make the gravity vary slightly over the surface. This is why planetary gravitational accelerations are generally given as an average value on the planet's equator.

Just as planets are not truly spheres, the orbits of solar system objects are not circular. Johannes Kepler, the prominent 17th-century German mathematician and astronomer, first realized that the orbits of planets are ellipses after analyzing a series of precise observations of the location of Mars that had been taken by his colleague, the distinguished Danish astronomer Tycho Brahe. Kepler drew rays from the Sun's center to the orbit of Mars and noted the date and time that Mars arrived on each of these rays. He noted that Mars swept out equal areas between itself and the Sun in equal times, and that Mars moved much faster when it was near the Sun than when it was farther from the Sun. Together, these observations convinced Kepler that the orbit was shaped as an ellipse and not as a circle, as had been previously assumed. Kepler defined three laws of orbital motion (listed in the table below), which he published in 1609 and 1619 in his books *New Astronomy* and *The Harmony of the World*. These three laws are still used as the basis for understanding orbits.

As Kepler observed, all orbits are ellipses, not circles. An ellipse can be thought of simply as a squashed circle, resembling an oval. The proper definition of an ellipse is the set of all points that have the same sum of distances to two given fixed points, called foci. To demonstrate this definition, take two pins, push them into a piece of stiff cardboard, and loop a string around the pins (see figure on page 11). The two pins are the foci of the ellipse. Pull the string away from the pins with a pencil, and draw the ellipse, keeping the string taut around the pins and the pencil all the way around. Adding the distance along the two string segments from the pencil to each of the pins will give the same answer each time: The ellipse is the set of all points that have the same sum of distances from the two foci.

KEPLER'S LAWS	
Kepler's first law:	A planet orbits the Sun following the path of an ellipse with the Sun at one focus.
Kepler's second law:	A line joining a planet to the Sun sweeps out equal areas in equal times (see figure on page 10).
Kepler's third law:	The closer a planet is to the Sun, the greater its speed. This is stated as: The square of the period of a planet T is proportional to the cube of its semimajor axis R, or $T \propto R^{\frac{3}{2}}$ as long as T is in years and R in AU.

Kepler's Second Law

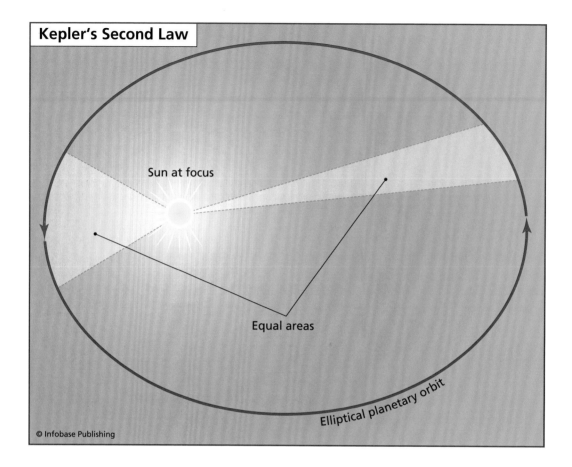

Sun at focus

Equal areas

Elliptical planetary orbit

© Infobase Publishing

Kepler's second law shows that the varying speed of a planet in its orbit requires that a line between the planet and the Sun sweep out equal areas in equal times.

The mathematical equation for an ellipse is

$$\frac{x^2}{a^2} + \frac{y^2}{b^2} = 1,$$

where x and y are the coordinates of all the points on the ellipse, and a and b are the semimajor and semiminor axes, respectively. The semimajor axis and semimajor axes would both be the radius if the shape was a circle, but two radii are needed for an ellipse. If a and b are equal, then the equation for the ellipse becomes the equation for a circle:

$$x^2 + y^2 = n,$$

where n is any constant.

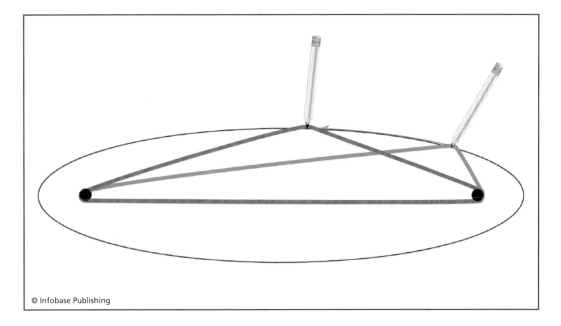

© Infobase Publishing

When drawing an ellipse with string and pins, it is obvious where the foci are (they are the pins). In the abstract, the foci can be calculated according to the following equations:

Coordinates of the first focus

$$= (+\sqrt{a^2 - b^2}, 0)$$

Coordinates of the second focus

$$= (-\sqrt{a^2 - b^2}, 0)$$

Making an ellipse with string and two pins— adding the distance along the two string segments from the pencil to each of the pins will give the same sum at every point around the ellipse. This method creates an ellipse with the pins at its foci.

In the case of an orbit the object being orbited (for example, the Sun) is located at one of the foci.

An important characteristic of an ellipse, perhaps the most important for orbital physics, is its eccentricity: a measure of how different are the semimajor and semiminor axes of the ellipse (see figure on page 12). Eccentricity is dimensionless and ranges from 0 to 1, where an eccentricity of zero means that the figure is a circle, and an eccentricity of 1 means that the ellipse has gone to its other extreme, a parabola (the reason an extreme ellipse becomes a parabola results from its

The semimajor and semiminor axes of an ellipse (or an orbit) are the elements used to calculate its eccentricity, and the body being orbited always lies at one of the foci.

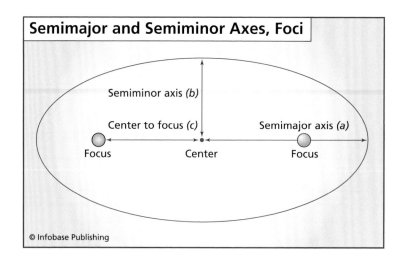

Semimajor and Semiminor Axes, Foci

Semiminor axis *(b)*

Center to focus *(c)*

Semimajor axis *(a)*

Focus

Center

Focus

© Infobase Publishing

definition as a conic section). One equation for eccentricity is as follows:

$$e = \sqrt{1 - \frac{b^2}{a^2}},$$

where *a* and *b* are the semimajor and semimajor axes, respectively. Another equation for eccentricity is

$$e = \frac{c}{a},$$

where *c* is the distance between the center of the ellipse and one focus. The eccentricities of the orbits of the planets vary widely, though most are very close to circles, as shown in the figure on page 13. Pluto has the most eccentric orbit at 0.244, and Mercury's orbit is also very eccentric, but the rest have eccentricities below 0.09.

While the characteristics of an ellipse drawn on a sheet of paper can be measured, orbits in space are more difficult to characterize. The ellipse itself has to be described, and then the ellipse's position in space, and then the motion of the body as it travels around the ellipse. Six parameters are needed to specify the motion of a body in its orbit and the position of the orbit. These are called the orbital elements (see the figure on page 15). The first three elements are used to determine where a body is in its orbit.

a **semimajor axis** The semimajor axis is half the width of the widest part of the orbit ellipse. For solar system bodies, the value of semimajor axis is typically expressed in units of AU. Mars's semimajor axis is 1.52366231 AU.

e **eccentricity** Eccentricity measures the amount by which an ellipse differs from a circle, as described above. An orbit with $e = 0$ is circular, and an orbit with $e = 1$ stretches into infinity and becomes a parabola. In between, the orbits are ellipses. The orbits of all large planets are almost circles: The Earth, for instance, has an eccentricity of 0.0068, and Mars's eccentricity is 0.09341233.

M **mean anomaly** Mean anomaly is an angle that moves in time from 0 to 360 degrees during one revolution, as if the planet were at the end of a hand of a clock and the Sun were at its center. This angle determines where in its orbit a planet is at a given time, and is defined to be 0 degrees at *perigee* (when

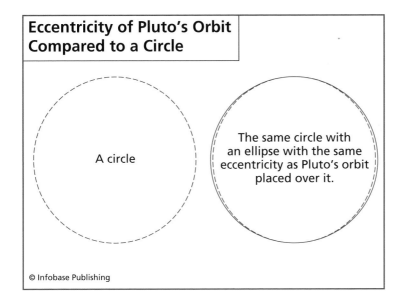

Eccentricity of Pluto's Orbit Compared to a Circle

A circle

The same circle with an ellipse with the same eccentricity as Pluto's orbit placed over it.

© Infobase Publishing

Though the orbits of planets are measurably eccentric, they deviate from circularity by very little. This figure shows the eccentricity of Pluto's orbits in comparison with a circle.

the planet is closest to the Sun) and 180 degrees at apogee (when the planet is farthest from the Sun). The equation for mean anomaly M is given as

$$M = M_0 + 360\left(\frac{t}{T}\right),$$

where M_0 is the value of M at time zero, T is the *orbital period*, and t is the time in question.

The next three Keplerian elements determine where the orbit is in space.

i **inclination**   For the case of a body orbiting the Sun, the inclination is the angle between the plane of the orbit of the body and the plane of the ecliptic (the plane in which the Earth's orbit lies). For the case of a body orbiting the Earth, the inclination is the angle between the plane of the body's orbit and the plane of the Earth's equator, such that an inclination of zero indicates that the body orbits directly over the equator, and an inclination of ninety indicates that the body orbits over the poles. If there is an orbital inclination greater than zero, then there is a line of intersection between the ecliptic plane and the orbital plane. This line is called the line of nodes. Mars's orbital inclination is 1.85061 degrees.

Ω **longitude of the ascending node**   After inclination is specified, there are still an infinite number of orbital planes possible: The line of nodes could cut through the Sun at any longitude around the Sun. Notice that the line of nodes emerges from the Sun in two places. One is called the ascending node (where the orbiting planet crosses the Sun's equator going from south to north). The other is called the descending node (where the orbiting planet crosses the Sun's equator going from north to south). Only one node needs to be specified, and by convention the ascending node is used. A second point in a planet's orbit is the vernal *equinox,* the spring day in which day and night have the same length ("equinox" means equal night), occurring where the plane of the planet's equator intersects its orbital plane. The angle between the vernal equinox γ and the ascending node *N* is called the longitude of the ascending node. Mars's longitude of the ascending node is 49.57854 degrees.

(ω) **argument of the perigee** The argument of the perigee is the angle (in the body's orbit plane) between the ascending node *N* and perihelion *P*, measured in the direction of the body's orbit. Mars's argument of the perigee is 336.04084 degrees.

The complexity of the six measurements shown above demonstrates the extreme attention to detail that is necessary when moving from simple theory ("every orbit is an

A series of parameters called orbital elements are used to describe exactly the orbit of a body.

Orbital Elements

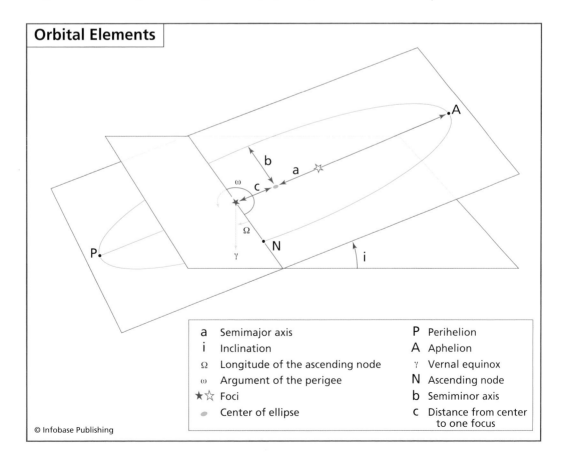

a	Semimajor axis	P	Perihelion
i	Inclination	A	Aphelion
Ω	Longitude of the ascending node	γ	Vernal equinox
ω	Argument of the perigee	N	Ascending node
★☆	Foci	b	Semiminor axis
●	Center of ellipse	C	Distance from center to one focus

© Infobase Publishing

ellipse") to measuring the movements of actual orbiting planets. Because of the gradual changes in orbits over time caused by gravitational interactions of many bodies and by changes within each planet natural orbits are complex, evolving motions. To plan with such accuracy space missions such as the recent Mars Exploration Rovers, each of which landed perfectly in their targets, just kilometers long on the surface of another planet, the mission planners must be masters of orbital parameters.

Seasons are created almost exclusively by the tilt of the planet's rotational axis, called its obliquity (see figure on page 17). While a planet rotates around the Sun, its axis always points in the same direction (the axis does wobble slightly, a movement called *precession*). The more extreme the obliquity, the more extreme will be the planet's seasons. The Earth's obliquity is not the most extreme in the solar system, as shown in the table on page 18.

Mars's obliquity, 25.2 degrees, is intermediate in the range of solar system values. The planet with the most extreme obliquity is Venus, with an obliquity of 177.3 degrees, followed by Pluto, with an obliquity of 122.5 degrees. An obliquity above 90 degrees means that the planet's north pole has passed through its orbital plane and now points south. This is similar to Uranus's state, with a rotational axis tipped until it almost lies flat in its orbital plane. With the exceptions of Mercury and Jupiter, therefore, all the planets have significant seasons caused by obliquity. When a planet with obliquity has its north pole tipped toward the Sun, the northern hemisphere receives more direct sunlight than the southern hemisphere does. The northern hemisphere then experiences summer, and the southern hemisphere is in winter. As the planet progresses in its orbit, revolving around the Sun until it has moved 180 degrees, then the southern hemisphere gets more direct sunlight, and the northern hemisphere is in winter. The more oblique the rotation axis, the more severe the seasons: In summer the hemisphere receives even more sunlight and the other hemisphere even less, and vice versa in winter. Summers are hotter and winters are colder. The obliquity of a planet may change over time, as well. Mars's obliquity may oscillate by

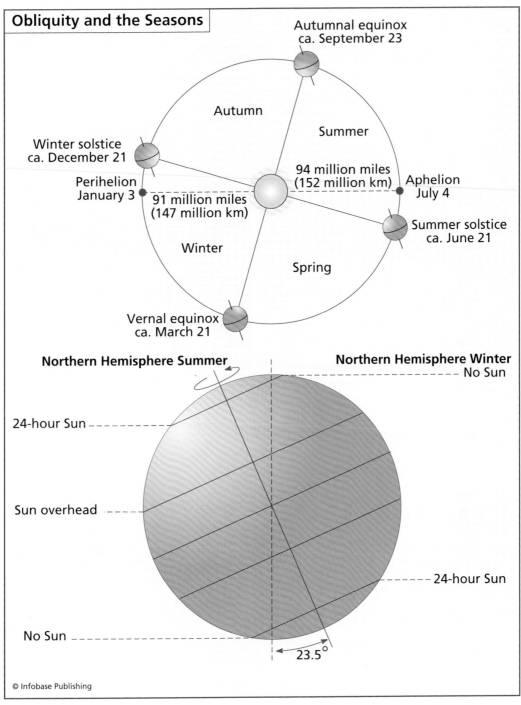

Obliquity and the Seasons

Autumnal equinox
ca. September 23

Autumn

Summer

Winter solstice
ca. December 21

94 million miles
(152 million km)

Aphelion
July 4

Perihelion
January 3

91 million miles
(147 million km)

Summer solstice
ca. June 21

Winter

Spring

Vernal equinox
ca. March 21

Northern Hemisphere Summer

Northern Hemisphere Winter
No Sun

24-hour Sun

Sun overhead

24-hour Sun

No Sun

23.5°

A planet's obliquity (the inclination of its equator to its orbital plane) is the primary cause of seasons.

OBLIQUITY, ORBITAL INCLINATION, AND ROTATIONAL DIRECTION FOR ALL THE PLANETS

Planet	Obliquity (inclination of the planet's equator to its orbit; tilt); remarkable values are in italic	Orbital inclination to the ecliptic (angle between the planet's orbital plane and the Earth's orbital plane); remarkable values are in italic	Rotational direction
Mercury	0° (though some scientists believe the planet is flipped over, so this value may be 180°)	7.01°	prograde
Venus	*177.3°*	3.39°	retrograde
Earth	23.45°	0° (by definition)	prograde
Mars	25.2°	1.85°	prograde
Jupiter	3.12°	1.30°	prograde
Saturn	26.73°	2.48°	prograde
Uranus	*97.6°*	0.77°	retrograde
Neptune	29.56°	1.77°	prograde
Pluto (now classified as a dwarf planet)	*122.5°*	*17.16°*	retrograde

as much as 20 degrees over time, creating seasons that are much more extreme. The Moon's stabilizing influence on the Earth has prevented large changes in obliquity and helped maintain a more constant climate, allowing life to continue and flourish.

If summer occurs when the Sun shines directly on the hemisphere in question, then the intensity of summer must depend in part on when it occurs in the planet's orbit. If the planet's axis tilts such that the hemisphere has summer at perihelion, when the planet is closest to the Sun, then it will be a much hotter summer than if that hemisphere had summer at aphelion, when the planet is farthest from the Sun (summer at the aphelion will also be shorter, since the planet is moving faster). It turns out that a planet's orbits do precess, that is, wobble, and so the positions of the seasons change over tens of thousands of years. This leads to a long-term cycle in seasonal severity.

In the northern hemisphere, the midpoint of summer occurs when the north pole points most directly toward the Sun. This is called the summer solstice, the longest day of the year, after which days become shorter. The northern hemisphere's winter solstice is its shortest day of the year. The reverse is true in the southern hemisphere. The planet's obliquity has the largest control over the severity of seasons, but there are secondary effects that also influence the temperature differences between seasons. In the summer the Sun is higher in the sky and so spends more time crossing the sky, and therefore the days are longer. This gives the Sun more time to heat the planet. In the winter the Sun is lower and the days are short, giving the Sun less time to heat the planet. Along with summer solstice and winter solstice, there are two other days that divide the year into quarters. The vernal equinox is the day in spring on which day and night are the same length (equinox means equal night). The autumnal equinox is the day in fall when day and night are the same length.

The distance of the Earth to the Sun has a very small effect on seasonal severity on Earth, because Earth's orbit is almost circular. Mars, however, has a much more eccentric orbit that influences its seasons considerably. Though its orbit will change over time, currently Mars is 1.65 AU from the Sun during its summer solstice and 1.38 AU from the Sun during its winter solstice. The northern hemisphere experiences cooler summers because it is farther from the Sun and therefore receives less sunlight, and it experiences warmer winters because it is relatively close to the Sun during its winter. The southern hemisphere, however, experiences far more severe seasons because of the eccentricity of Mars's orbit. Its winters occur when Mars is far from the Sun, and its summers occur when Mars is close to the Sun. When Mars is close to the Sun it is also moving faster, and so the southern hemisphere summers are also shorter. Because of its particular combination of obliquity and orbital eccentricity, Mars receives significantly more heat from the Sun during the southern hemisphere summer than it does during the northern hemisphere summer. The total heat received by the planet is thus not even over the year: Mars receives far more energy during southern hemisphere summer than it does during any other part of the year. This unevenness drives Mars's yearly cycle of carbon dioxide freezing and sublimation: The polar ice caps grow significantly during northern hemisphere summer such that the loss of carbon dioxide in the atmosphere actually causes a drop in atmospheric pressure. During southern hemisphere summer

The Mars Global Surveyor camera captured these dust storms in the Martian northern spring. (NASA/ PL/Malin Space Science Systems)

the ice caps sublimate (transform from solid to gas) back into the atmosphere and raise pressure again. Seasonal variation on Mars drives not just the freezing and sublimation of its ice caps but also the birth and death of planet-encompassing dust storms. Temperature contrasts between the frozen carbon dioxide polar cap and the warm ground adjacent to it, combined with a flow of cool polar air evaporating off the cap, form dust storms at high latitudes. The dust storms shown on page 20 were photographed by the *Mars Global Surveyor* Mars Orbiter Camera in May 2002.

Mars's orbit and its changes with time are newly and precisely modeled with data from the most recent Mars missions. The ability to model not only the seasonal climate changes on the planet today but the severity of seasons in the past makes possible a new link between surface geologists and climate modelers: Martian surface geologists see evidence for glaciation on the planet in the past, and this finding is in agreement with new information about the severity of Mars's winters during periods of more extreme obliquity. A planet's orbit and obliquity are intimately tied to its climate, and the planet's climate is intimately tied to its landforms. As study on Mars progresses, all these fields move together to create a cohesive understanding of the planet as a whole.

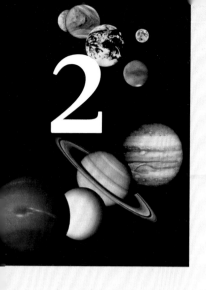

Planetary Evolution and the Martian Interior

The majority of the data that can be gathered about another planet concerns only its surface, since the majority of planetary data consists of photos and other kinds of remote sensing. Both orbiters and surface landers mainly measure the appearance, temperature, dynamics, and composition of the surface. In addition to investigating the surface of the planet, scientists would like to know the composition and temperature of the interior. Even on Earth people cannot drill deeply enough to measure the interior directly (the deepest drill hole has reached only 7.5 miles [12.3 km]). Scientists have developed ways, therefore, to estimate the interior composition and temperature from other measurements.

The interior of *terrestrial planets* is commonly divided into the crust, *lithosphere*, mantle, and *core*, as described in the sidebar "Interior Structure of the Terrestrial Planets" on page 23. On the Earth, researchers can comb the surface for pieces of the interior that have been carried up in volcanic eruptions and have built up huge databases on the compositions of magma that are a fairly direct way to learn about the interior from which they melted. For Mars there are only a relative handful of Martian meteorites (meteorites that originated on Mars and passed through space to Earth) to help us under-

INTERIOR STRUCTURE OF THE TERRESTRIAL PLANETS

The outermost layer of the terrestrial planets is the *crust,* made up of one or more of the following: igneous rocks (rocks that were once hot enough to be completely molten), metamorphic rocks (rocks that have been changed from their original state by heat or pressure, but were never liquid), and sedimentary rocks (rocks that are made of mineral grains that were transported by water or air). Sedimentary rocks, such as shale and limestone, can only be made at the surface of the planet, by surface processes such as rivers, glaciers, oceans, and wind. Sedimentary rocks make up the majority of rock outcrops at the surface of the Earth, and they may also exist on Mars. Igneous rocks make up the majority of the crust on Mars and on Venus, and probably on Mercury, though its surface is not well known because of the difficulty in taking good images of a body so close to the Sun. Igneous rocks predominate on Venus and probably Mercury because water is the main driving force for the formation of sedimentary rocks, and those planets are dry. Mars alone has sedimentary rocks similar to those on Earth. *Plate tectonics* on Earth is the main process that creates metamorphic rocks, which are commonly made by burial under growing mountains as tectonic plates collide or drive under each other. Since neither surface water nor plate tectonics exist on Venus or Mercury, they would be expected to be covered almost exclusively with igneous rocks.

Radioactive decay of *atoms* throughout the Earth and other planets produces heat, and the interior of the Earth and most other planets is hotter than the surface. Heat radiates away from the planet into space, and the planet is cooling through time. If the planet had only the heat from its initial formation, it can be calculated that planets the size of Earth and smaller, including Mars, would have completely cooled by this time. Scientists can measure heat flowing out of the Earth, however, by placing thermal measuring devices in holes dug deeply into the soil, and they know that heat flux (the amount of heat moving through a unit of surface area in a unit of time) is different in different parts of the world. Heat flux in areas of active volcanism, for example, is higher than heat flux in quiet areas. It is possible to calculate the likely internal temperatures of the Earth based on heat flux at the surface. This leads to the conclusion that the Earth's shallow interior, not much deeper than the hard, cold crust mankind lives upon, is at 2,190 to 2,640°F (1,200 to 1,450°C). Similar calculations can be made for other planets remotely, but without the precision of the answers for the Earth.

(continues)

(continued)

Rocks at temperatures like those in Earth's mantle are able to flow over geologic time. They are not liquid, but there is enough energy in the atoms of the crystals that the crystals can deform and "creep" in response to pressure, given thousands or millions of years. (For more, see the sidebar "Rheology, or How Solids Can Flow" on page 79.) The interior of the Earth is moving in this way, in giant circulation patterns driven by heat escaping toward the surface and radiating into space. This movement in response to heat is called *convection*.

Plate tectonics, the movement of the brittle outside of the Earth, is caused in part by these internal convective movements, but largely by the pull of oceanic plates descending into the mantle at *subduction zones*. At the surface, this movement is only one or two inches (a few centimeters) a year. Scientists are constantly looking for evidence for plate tectonics on other planets and, for the most part, not finding it; apparently other terrestrial planets have not experienced plate tectonics, likely because they have insufficient surface water, which lubricates plate movement and allows the mantle to flow more easily. There may be some ice plate tectonics on Europa, but there is nothing in the solar system like the complex plate configurations and movements on the Earth. On Mars there may be some evidence for plate tectonics that ended early in the planet's history, left recorded in a few regions of the crust that have retained a magnetic field. Mars and the other terrestrial planets and the Moon are all considered *one-plate planets*, planets where their crust and upper mantle have cooled into a stiff shell that moves as a unit. Without plate tectonics, there are no volcanic arcs such as Japan or the Cascades, and there are no *mid-ocean ridges* at which new oceanic crust is produced. Surface features on one-plate planets are therefore different from those on Earth.

Below the crust and above the core, the planet's material is called the mantle. The uppermost mantle is too cool to be able to flow, except over many millions of years, and so it moves as a unit with the crust. Together, these two cool, connected layers stand its interior. In the future space missions may achieve sample return from Mars, as they did from the Moon, but that is a distant thought at the moment. On the Earth scientists can measure *seismic waves* and use their changes in velocity to plot out the surfaces inside the Earth at which composition or mineralogy changes, such as at the interface between the lithosphere and mantle. On Mars this is not yet an option. In the future manned or even unmanned missions may place

are called the lithosphere. Beneath the lithosphere, the remaining mantle might be hot enough to flow. This is certainly true on the Earth and is probably the case on the other terrestrial planets. The mantles of terrestrial planets are thought to be mainly made of minerals based on silicon atoms. The most common minerals at shallow depths in the mantle are *olivine* (also known as the semiprecious gem peridot) and pyroxene at shallow depths, which then convert to other minerals at the higher pressures of the deep interior. Mantles on other differentiated terrestrial planets are thought to be similar.

Based on an analysis of the bulk "silicate Earth" (the mantle and crust, made mostly of minerals based on silicon atoms) compared to the composition of primitive meteorites that represent the material the inner planets were made of, the silicate Earth is clearly missing a lot of iron and some nickel. Models of planetary formation also show that the heat of *accretion* (the initial formation of the planet) will cause iron to melt and sink into the deep interior of the Earth. The core of the Earth, then, is made of iron with some nickel and a few percent of other *elements*. The other terrestrial planets almost certainly have iron cores, and even the Moon probably has a tiny iron core with a radius of a few hundred kilometers at most.

This is the structure of the Earth, used as the starting point in understanding the structures of other terrestrial planets: The outermost cool, thin veneer of the Earth is the crust. The crust is coupled to the coolest, uppermost mantle, and together they are called the lithosphere. Under the lithosphere is the convecting mantle, and beneath that, the core. The outer core is liquid metal, and the inner core is solid metal.

How the crusts formed seems to differ among the planets, as does the composition of the mantle (though thought to be always silicate) and core (though thought to be always iron-dominated), and the heat and convective activity of the mantle. Theorizing about the degree of these differences and the reasons for their existence is a large part of planetary geology.

seismometers on Mars that can radio seismic wave measurements back to Earth (these exist on the Moon), but for now, only the unmanned surface rovers and remote sensing from orbiters provide data on the planet. Despite this discouraging lack of data, there are a number of clever ways to infer the composition and structure of the Martian interior and even to hypothesize about Mars's past. This chapter will discuss what is known and what is inferred about the Martian interior, and what scientists speculate about the early formation of Mars that allowed it to evolve into the planet it is today.

DIRECT OBSERVATIONS OF THE COMPOSITION OF MARS

Spectra giving compositional information have been obtained from Earth-based and orbital-based observations and from measurements made by the Viking and Pathfinder landers. These sources give information only on the composition of the crust, and that information is limited. Some of the best ideas on the composition of Mars come from what is known about the composition of the Earth along with the assumption that since the two bodies formed near each other in the solar system they should be composed of similar materials; this is an indirect method for learning about Mars. The best direct data on the composition of Mars comes from Martian meteorites. Several types of meteorites have been shown to have come to Earth from Mars. These are meteorites previously named for the places on Earth they were found: Shergottites, Nakhlites, and Chassignites. Collectively they are called SNC (popularly pronounced "snick") meteorites.

The first evidence that these meteorites are from Mars is their young crystallization ages. When rated using radioactive *isotopes,* these meteorites were found to have cooled and crystallized at times ranging from 150 million to 1.3 billion years ago. (See the sidebars "Determining Age from Radioactive Isotopes," on page 39, and "Elements and Isotopes," on page 27.) Only a large planet could have remained volcanically active for so long after its formation at 4.56 billion years ago. These meteorites, therefore, did not come from the asteroid belt and were not related to chondrites and the other primitive meteorites. Later, the gases from bubbles trapped in the meteorites were carefully extracted and analyzed, and it was found that their compositions and isotopic ratios matched those from the Martian atmosphere measured by the Viking lander and in no way resembled Earth's atmosphere. These gas bubbles are believed to have been formed in the meteorites by the shock of the impact that ejected them from Mars; each of the

ELEMENTS AND ISOTOPES

All the materials in the solar system are made of atoms or of parts of atoms. A family of atoms that all have the same number of positively charged particles in their nuclei (the center of the atom) is called an element: Oxygen and iron are elements, as are aluminum, helium, carbon, silicon, platinum, gold, hydrogen, and well over 200 others. Every single atom of oxygen has eight positively charged particles, called protons, in its nucleus. The number of protons in an atom's nucleus is called its *atomic number:* All oxygen atoms have an atomic number of 8, and that is what makes them all oxygen atoms.

Naturally occurring nonradioactive oxygen, however, can have either eight, nine, or 10 uncharged particles, called neutrons, in its nucleus, as well. Different weights of the same element caused by addition of neutrons are called isotopes. The sum of the protons and neutrons in an atom's nucleus is called its *mass number.* Oxygen can have mass numbers of 16 (eight positively charged particles and eight uncharged particles), 17 (eight protons and nine neutrons), or 18 (eight protons and 10 neutrons). These isotopes are written as ^{16}O, ^{17}O, and ^{18}O. The first, ^{16}O, is by far the most common of the three isotopes of oxygen.

Atoms, regardless of their isotope, combine together to make molecules and compounds. For example, carbon (C) and hydrogen (H) molecules combine to make methane, a common gas constituent of the outer planets. Methane consists of one carbon atom and four hydrogen atoms and is shown symbolically as CH_4. Whenever a subscript is placed by the symbol of an element, it indicates how many of those atoms go into the makeup of that molecule or compound.

Quantities of elements in the various planets and moons, and ratios of isotopes, are important ways to determine whether the planets and moons formed from the same material or different materials. Oxygen again is a good example. If quantities of each of the oxygen isotopes are measured in every rock on Earth and a graph is made of the ratios of $^{17}O/^{16}O$ versus $^{18}O/^{16}O$, the points on the graph will form a line with a certain slope (the slope is 1/2, in fact). The fact that the data forms a line means that the material that formed the Earth was homogeneous; beyond rocks, the oxygen isotopes in every living thing and in the atmosphere also lie on this slope. The materials on the Moon also show this same slope. By measuring oxygen isotopes in many different kinds of solar system materials, it has now been shown that the slope of the plot $^{17}O/^{16}O$

(continues)

(continued)

versus $^{18}O/^{16}O$ is one-half for every object, but each object's line is offset from the others by some amount. Each solar system object lies along a different parallel line.

At first it was thought that the distribution of oxygen isotopes in the solar system was determined by their mass: The more massive isotopes stayed closer to the huge gravitational force of the Sun, and the lighter isotopes strayed farther out into the solar system. Studies of very primitive meteorites called chondrites, thought to be the most primitive, early material in the solar system, showed to the contrary that they have heterogeneous oxygen isotope ratios, and therefore oxygen isotopes were not evenly spread in the early solar system. Scientists then recognized that temperature also affects oxygen isotopic ratios: At different temperatures, different ratios of oxygen isotopes condense. As material in the early solar system cooled, it is thought that first aluminum oxide condensed, at a temperature of about 2,440°F (1,340°C), and then calcium-titanium oxide ($CaTiO_3$), at a temperature of about 2,300°F (1,260°C), and then a calcium-aluminum-silicon-oxide ($Ca_2Al_2SiO_7$), at a temperature of about 2,200°F (1,210°C), and so on through other compounds down to iron-nickel alloy at 1,800°F (990°C) and water, at −165°F (−110°C) (this low temperature for the condensation of water is caused by the very low pressure of space). Since oxygen isotopic ratios vary with temperature, each of these oxides would have a slightly different isotopic ratio, even if they came from the same place in the solar system.

The key process that determines the oxygen isotopes available at different points in the early solar system nebula seems to be that simple compounds created with ^{18}O

Martian meteorites had to have been ejected from Mars as a splash from a large impactor striking Mars. Mathematical calculations show that the chances of *ejecta* from Mars falling to Earth are high, and the time of passage between the two planets can be short; in fact, it is likely that many pieces make the transit from Mars to the Earth in less than two million years. Based on damage done to the surface of the meteorites by cosmic rays while they were passing through space, the SNC meteorites found on Earth seem to have been ejected in one or more impacts on Mars within the last one million to 20 million years.

New Martian meteorites are identified almost every year, now that scientists know they are present and meteorite

are relatively stable at high temperatures, while those made with the other two isotopes break down more easily and at lower temperatures. Some scientists therefore think that ^{17}O and ^{18}O were concentrated in the middle of the nebular cloud, and ^{16}O was more common at the edge. Despite these details, though, the basic fact remains true: Each solar system body has its own slope on the graph of oxygen isotope ratios.

Most atoms are stable. A carbon-12 atom, for example, remains a carbon-12 atom forever, and an oxygen-16 atom remains an oxygen-16 atom forever, but certain atoms eventually disintegrate into a totally new atom. These atoms are said to be "unstable" or "radioactive." An unstable atom has excess internal energy, with the result that the nucleus can undergo a spontaneous change toward a more stable form. This is called "radioactive decay." Unstable isotopes (radioactive isotopes) are called "radioisotopes." Some elements, such as uranium, have no stable isotopes. The rate at which unstable elements decay is measured as a "half-life," the time it takes for half of the unstable atoms to have decayed. After one half-life, half the unstable atoms remain; after two half-lives, one-quarter remain, and so forth. Half-lives vary from parts of a second to millions of years, depending on the atom being considered. Whenever an isotope decays, it gives off energy, which can heat and also damage the material around it. Decay of radioisotopes is a major source of the internal heat of the Earth today: The heat generated by accreting the Earth out of smaller bodies and the heat generated by the giant impactor that formed the Moon have long since conducted away into space.

collecting trips are common. There are 49 known Martian meteorites, listed in the table on page 33. They have varying importance in the eyes of the scientific community. All the SNC meteorites are igneous rocks, either high-magnesium lavas or rocks derived from high-magnesium lavas. This makes them useful for inferring information about the Martian mantle, where the source regions that melted to create these rocks must have been. One important Martian meteorite, ALH 84001, is estimated to have taken about 3 million years to make it from Mars to Earth, moving through space on an indirect path.

Curt Mileikowsky, a scientist from the Royal Academy in Stockholm, and his colleagues calculated that, over the age of

EETA79001

The EETA 79001 meteorite provided the first strong proof that meteorites could come from Mars. The Mars Exploration Rover Opportunity has discovered that a rock called "Bounce" at Meridiani Planum has a very similar mineral composition to this meteorite and likely shares common origins. Bounce itself is thought to have originated outside the area surrounding Opportunity's landing site; an impact or collision likely threw the rock away from its primary home, just as an impact threw the meteorite into space, eventually to land on Earth in Antarctica. The side of the cube at the lower left in this image measures ca. 0.4 inches (1 cm). (NASA/JSC/JPL/Lunar Planetary Institute)

the solar system, a billion tons of material has been ejected from Mars by meteorite impacts and come to Earth. There is, therefore, very little of this material. Mileikowsky estimates that every 100,000 years a rock makes it straight from Mars to Earth in under a year and, on the strength of this argument, suggests that if Mars developed life its transfer to Earth was likely.

The Martian meteorite EETA 79001, named for the Elephant Moraine location of its find in 1979 in Antarctica, is a basaltic rock that melted from the Martian interior and so holds important clues about the composition of Mars. The image of EETA 79001 here shows a sawn face of the rock with gray, fine-grained minerals and black areas of glass. The glass in EETA 79001 contained the first gas bubbles analyzed that showed the clear signature of the Martian atmosphere and thus proved that the meteorite came from Mars.

Meteorites are commonly found in deserts or in the Antarctic, not because they fall there with any more frequency than anywhere else, but because in deserts and on ice sheets stray rocks are easier to spot and are much more likely to be meteorites than anything else. In the deserts of northern Africa, nomadic tribes collect rocks and bring them to local dealers, who sort them and sell them to meteorite dealers from other continents. These dealers are increasingly aware of the possibility of finding Martian meteorites, and, after analysis, new Martian meteorite finds are frequently announced by scientific groups from around the world.

Private meteorite collections, however, are popular, and eager buyers from the private sector are driving up prices beyond what academics can afford. University researchers have often agreed to classify meteorites for dealers in return for a portion of the meteorite, usually two g or 40 mass percent (a percent of the mass of the sample, as contrasted with percent of the volume of the sample or any other way of measuring the sample), whichever is larger. Some dealers are now learning to make their own classifications, and so university scientists have far less access to the meteorites found in deserts in Africa or other non-Antarctic areas. (Classification is identifying the meteorite class; for example, ordinary chondrite or achondrite, and to further classify the sample within the class.)

The main means scientists now have to obtain new meteorites is from their own Antarctic meteorite searches. Several countries run them and often share results. The American effort, the Antarctic Search for Meteorites (ANSMET), is a program that sends scientists and other interested participants to Antarctica every fall, where they spend about six weeks searching for meteorites on the ice. This program is the only regular, reliable source for new meteorites for the scientific community, and it has been running since 1976, funded by the U.S. National Science Foundation (NSF). Over the years the program has returned more than 10,000 meteorites.

Ralph Harvey, the principal investigator of the project, has to choose a team for the year's search. He picks candidates from paper application letters he receives (not accepting e-mail applications has significantly cut down on applicants, and therefore on his workload). These candidates are almost all graduate students or established scientists. Between six and 12 people are chosen for the team, and in addition Ralph goes every year.

The team flies to Christchurch, New Zealand, in November, just before Thanksgiving. At the NSF Office of Polar Programs (OPP), all participants are given the necessary clothing. The team then gets on an uncomfortable National Guard plane and makes the six-hour flight to McMurdo Base, on Antarctica. After a week of training, they take off onto the ice with snowmobiles for six weeks of camping and collecting meteorites.

Each meteorite found on the ice is photographed in place, packaged individually with a field number and a brief description, and kept frozen. For some time any sample that looked special in the field would be given a special field number—that is how ALH 84001 obtained its important "001" number. All the meteorites are shipped on a freezer ship to California and then driven in a freezer truck to the Johnson Space Center freezers. At Johnson Space Center, finally, they are thawed, but only using a freeze-dry process, so the sample never soaks in terrestrial water. Getting the meteorites back to the Johnson Space Center takes about four months, and after freeze-drying a chip of each sample is sent to the Smithsonian to be classified. The most interesting-looking material comes first, usually by the summer following the search.

ANSMET, though originally funded by the NSF alone, is really a cooperative process between the National Aeronautics and Space Administration (NASA), which curates the bulk of the meteorites, the NSF, and the Smithsonian, who work in concert to run the searches, process and classify the meteorites, and make them available to the scientific community.

The team members always find travel and life on Antarctica an intense experience. On average, the team members find

(continues on page 36)

KNOWN MARTIAN METEORITES AS OF MID-2009

Meteorite Name	Location Found	Date Found or Purchased from Local Dealer	Mass (g)	Type
Chassigny	France	October 3, 1815	~4,000	dunite
Shergotty	India	August 25, 1865	~5,000	*basalt*
Nakhla	Egypt	June 28, 1911	~10,000	clinopyroxenite
Lafayette	Indiana, USA	Before 1931	~800	clinopyroxenite
Governador Valadares	Brazil	1958	158	clinopyroxenite
Zagami	Nigeria	October 3, 1962	~18,000	basalt
Allan Hills (ALH) A77005	Antarctica	December 29, 1977	482	peridotite
Yamato 793605	Antarctica	1979	16	peridotite
Elephant Moraine (EET) A79001	Antarctica	January 13, 1980	7,900	basalt
Allan Hills 84001	Antarctica	December 27, 1984	1,939.9	orthopyroxenite
Lewis Cliff 88516	Antarctica	December 22, 1988	13.2	peridotite
Queen Alexandra Range (QUE) 94201	Antarctica	December 16, 1994	12.0	basalt
Dar al Gani 735	Libya	1996–1997	588	basalt
Dar al Gani 489	Libya	1997	2,146	basalt
Dar al Gani 670	Libya	1998–1999	1,619	basalt
Dar al Gani 476	Libya	May 1, 1998	2,015	basalt
Dar al Gani 876	Libya	May 7, 1998	6.2	basalt
Dar al Gani 975	Libya	August 21, 1999	27.55	basalt
Dar al Gani 1037	Libya	1999	4,012.4	basalt

(continues)

KNOWN MARTIAN METEORITES AS OF MID-2009
(continued)

Meteorite Name	Location Found	Date Found or Purchased from Local Dealer	Mass (g)	Type
Yamato 980459	Antarctica, Yamato Mtns	December 4, 1998	82.46	basalt
Los Angeles 001	California, USA	October 30, 1999	452.6	basalt
Sayh al Uhaymir 005	Oman	November 26, 1999	1,344	basalt
Sayh al Uhaymir 008	Oman	November 26, 1999	8,579	basalt
Sayh al Uhaymir 051	Oman	August 1, 2000	436	basalt
Sayh al Uhaymir 094	Oman	February 8, 2001	233.3	basalt
Sayh al Uhaymir 060	Oman	June 27, 2001	42.28	basalt
Sayh al Uhaymir 090	Oman	January 19, 2002	94.84	basalt
Sayh al Uhaymir 120	Oman	November 17, 2002	75	basalt
Sayh al Uhaymir 150	Oman	October 8, 2002	107.7	basalt
Sayh al Uhaymir 125	Oman	November 19, 2003	31.7	basalt
Sayh al Uhaymir 130	Oman	January 11, 2004	278.5	basalt
Sayh al Uhaymir 131	Oman	January 11, 2004	168	
Dhofar 019	Oman	January 24, 2000	1,056	basalt
Grove Mountains 99027	Antarctica	February 8, 2000	9.97	peridotite
Dhofar 378	Oman	June 17, 2000	15	basalt
Northwest Africa 2737	Morocco	August, 2000	611	dunite
Northwest Africa 480	Morocco	November 2000	28	basalt
Northwest Africa 1460	Morocco	November 2001	70.2	basalt

Meteorite Name	Location Found	Date Found or Purchased from Local Dealer	Mass (g)	Type
Yamato 000593	Antarctica	November 29, 2000	13,700	clinopyroxenite
Yamato 000749	Antarctica	December 3, 2000	1,300	clinopyroxenite
Yamato 000802	Antarctica	January 2003?	22	clinopyroxenite
Yamato 000027	Antarctica	November 2000	9.68	peridotite
Yamato 000047	Antarctica	November 2000	5.34	peridotite
Yamato 000097	Antarctica	November 2000	24.48	peridotite
Northwest Africa 817	Morocco	December 2000	104	clinopyroxenite
Northwest Africa 4797	Morocco	2001	15	peridotite
Northwest Africa 1669	Morocco	January 2001	35.85	basalt
Northwest Africa 1950	Morocco	January 2001	797	peridotite
Northwest Africa 856	Morocco	March 2001	320	basalt
Northwest Africa 1068	Morocco	April 2001	654	basalt
Northwest Africa 1110	Morocco	January 2002	118	basalt
Northwest Africa 1775	Morocco	2002	25	basalt
Northwest Africa 2373	Morocco	August 2004	18.1	basalt
Northwest Africa 998	Algeria or Morocco	September 2001	456	clinopyroxenite
YA1075	Antarctica	before 2002	55	peridotite
Northwest Africa 1195	Morocco	March 2002	315	basalt
Grove Mountains 020090	Antarctica	January 4, 2003	7.5	peridotite
Northwest Africa 5029	Morocco	March 2003	14.7	basalt
Northwest Africa 2046	Algeria	September 2003	63	basalt

(continues)

KNOWN MARTIAN METEORITES AS OF MID-2009
(continued)

Meteorite Name	Location Found	Date Found or Purchased from Local Dealer	Mass (g)	Type
MIL 03346	Antarctica, Miller Range	December 15, 2003	715.2	clinopyroxenite
Roberts Massif 04261	Antarctica	2004	78.8	basalt
Roberts Massif 04262	Antarctica	2004	204.6	basalt
Northwest Africa 3171	Algeria	February 2004	506	basalt
Northwest Africa 2626	Algeria	November 2004	31.07	basalt
Northwest Africa 2646	Algeria or Morocco	December 2004	9.3	clinopyroxenite
Northwest Africa 2969	Morocco	August 2005	11.7	basalt
Northwest Africa 2975	Algeria	November 2005	70.1	basalt
Larkman Nunatak 06319	Antarctica	2006	78.6	basalt
Northwest Africa 4468	NW Africa	Summer 2006	675	basalt
Northwest Africa 4480	Morocco	Summer 2006	13	basalt
Northwest Africa 4527	Algeria	Summer 2006	10.1	basalt
Northwest Africa 2800	Morocco	July 2007	686	basalt
Northwest Africa 2800	Morocco	2007	686	basalt
Northwest Africa 2990	Morocco	2007	0.363	basalt

Notes: Entries with borders indicate they are fragments of the same meteorite; meteorites are organized by the date of finding the first piece of a given meteorite.

(continued from page 32)

about 300 meteorites in a season, though in one recent year they found almost 1,500. NASA has also funded an additional four people to be a reconnaissance team, to move ahead of the main team and guide them to the richest areas.

Life on the ice for the teams during searches is a little more basic. Two people stay in each tent, partly for heat, and partly for safety. The tents are heated with small camping stoves, and as the season wears on the stoves can begin to malfunction. It is not uncommon to see, while walking through camp, a tent open suddenly and a flaming stove soar out into the snow. The deliberate pace of operations and the vagaries of weather and equipment demand a lot of patience from the participants. Cari Corrigan, a geologist with the Smithsonian Institution and member of the permanent ANSMET team, explains that team members learn to accept things that would otherwise be miserable or frustrating and learn to accept delays without irritation.

MAKING MARS: COMPOSITIONS AND PROCESSES

Since the planets formed early in the history of the solar system, they largely formed out of primitive, unprocessed material from the solar nebula. Today the best analogs to primitive solar system material available are meteorites. Which meteorite composition represents the most primitive material can be inferred by comparing the elements in the meteorite with the elements that make up the Sun. The Sun, since it makes up more than 99 percent of the material in the solar system, is probably a good measure of an average solar system composition. Other materials have been processed by being smashed by other meteorites, by being partly melted and having the solid and melted portions separated (differentiation), perhaps repeatedly, or by being mixed and heated with water. Thus, material from planets, and even most material from asteroids and meteorites, does not represent the initial bulk composition of the solar system.

The composition of the Sun can be determined from the spectra of electromagnetic radiation that the Sun emits. Each kind of atom can only absorb certain quantities of energy, called quanta. When an atom absorbs energy one or more of its orbiting electrons may be forced to move away from the atomic nucleus and begin orbiting at a higher level. Empty orbitals are

One class of meteorites, the carbonaceous chondrites, have compositions almost exactly like the Sun, as shown in the plot, and so are thought to represent the bulk composition of the planetary disk that formed the planets (modeled after Ringwood, 1979).

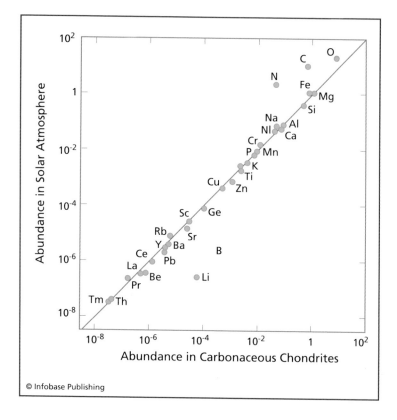

inherently unstable, so the energized atom will eventually emit a specific quanta of energy and allow its electrons to return to their stable orbitals. The quantas of energy an atom emits are characteristic of the type of element it is. When light from the Sun is split into a spectrum showing the intensity of each wavelength, the peaks in intensity correspond to the particular elements in the Sun that are emitting energy. (For more, see the sidebar "Remote Sensing" on page 102.) By studying these spectra, the composition of the Sun has been carefully determined. Is this composition accurate for the whole Sun? No, it only reflects the composition of the part of the Sun that is emitting the radiation. The solar composition is thought to be reasonably accurate for the bulk Sun, but in fact it is an estimate.

A class of meteorites called chondrites have compositions very similar to the Sun's, except in the very *volatile* elements like helium and hydrogen. If the abundance of elements in the

(continues on page 42)

DETERMINING AGE FROM RADIOACTIVE ISOTOPES

Each element exists in the form of atoms with several different-sized nuclei, called isotopes. Consider the element carbon. All carbon atoms have six protons in their nuclei, but they can have different numbers of neutrons. Protons determine the kind of element the atom is because protons have a positive charge, and to balance their positive charge, the atom has negatively charged electrons orbiting around its nucleus. It is the structure and number of electrons and the size of the atom that determine how it interacts with other atoms, and thus makes all atoms with the same number of protons act alike. Neutrons, on the other hand, make an atom heavier, but do not change its chemical interactions very much. The atoms of an element that have different numbers of neutrons are called isotopes. Carbon has three isotopes, with atomic masses 12, 13, and 14. They are denoted ^{12}C, ^{13}C, and ^{14}C.

Most atoms are stable. A ^{12}C atom, for example, remains a ^{12}C atom forever, and an ^{16}O (oxygen) atom remains an ^{16}O atom forever, but certain atoms eventually disintegrate into a totally new atom. These atoms are said to be unstable or radioactive. An unstable atom has excess internal energy, with the result that the nucleus can undergo a spontaneous change toward a more stable form. This is called radioactive decay.

Unstable isotopes (which are thus radioactive) are called radioisotopes. Some elements, such as uranium, have no stable isotopes. Other elements have no radioactive isotopes. The rate at which unstable elements decay is measured as a *half-life*, the time it takes for half of the unstable atoms to have decayed. After one half-life, half the unstable atoms remain; after two half-lives, one-quarter remain, and so forth. Half-lives vary from parts of a second to billions of years, depending on the atom being considered. The radioactive element is called the parent, and the product of decay is called the daughter.

When rocks form, their crystals contain some amount of radioactive isotopes. Different crystals have differently sized spaces in their lattices, so some minerals are more likely to incorporate certain isotopes than others. The mineral zircon, for example, usually contains a measurable amount of radioactive lead (atomic abbreviation Pb). When the crystal forms, it contains some ratio of parent and daughter atoms. As time passes, the parent atoms continue to decay according to the rate given by their half-life, and the population of daughter atoms in the crystal increases. By measuring the concentrations of parent and daughter atoms, the age of the rock can be determined.

(continues)

(continued)

To learn the math to calculate the age of materials based on radioactive decay, read this section. Otherwise, skip to the end of the math section to learn about the ages of objects in the solar system.

Consider the case of the radioactive decay system ^{87}Rb (rubidium). It decays to ^{87}Sr (strontium) with a half-life of 49 billion years. In a given crystal, the amount of ^{87}Sr existing now is equal to the original ^{87}Sr that was incorporated in the crystal when it formed, plus the amount of ^{87}Rb that has decayed since the crystal formed. This can be written mathematically as:

$$^{87}Sr_{now} = {}^{87}Sr_{original} + ({}^{87}Rb_{original} - {}^{87}Rb_{now}).$$

The amount of rubidium now is related to the original amount by its rate of decay. This can be expressed in a simple relationship that shows that the change in the number of parent atoms n is equal to the original number n_o times one over the rate of decay, called λ (the equations will now be generalized for use with any isotope system):

$$\frac{-dn}{dt} = \lambda n_o,$$

where dn means the change in n, the original number of atoms, and dt means the change in time. To get the number of atoms present now, the expression needs to be rearranged so that the time terms are on one side, and the n terms on the other, and then integrated, and the final result is:

$$n_0 = ne^{\lambda t}.$$

The number of daughter atoms formed by decay, D, is equal to the number of parent atoms that decayed:

$$D = n_0 - n.$$

Also, from the previous equation, $n_o = ne^{\lambda t}$. That expression can be substituted into the equation for D in order to remove the term n_o:

$$D = n(e^{\lambda t} - 1).$$

Then, finally, if the number of daughter atoms when the system began was D_o, then the number of daughter atoms now is

$$D = D_0 + n(e^{\lambda t} - 1).$$

This is the equation that allows geologists to determine the age of materials based on radiogenic systems. The material in question is ground up, dissolved in acid, and vaporized in an instrument called a mass spectrometer. The mass spectrometer measures the relative abundances of the isotopes in question, and then the time over which they have been decaying can be calculated.

The values of D and n are measured for a number of minerals in the same rock, or a number of rocks from the same outcrop, and the data is plotted on a graph of D v. n (often D and n are measured as ratios of some stable isotope, simply because it is easier for the mass spectrometer to measure ratios accurately than it is to measure absolute abundances). The slope of the line the data forms is $e^{\lambda t} - 1$. This relation can be solved for t, the time since the rocks formed. This technique also neatly gets around the problem of knowing D_o, the initial concentration of daughter isotopes: D_o ends up being the y-intercept of the graph.

Radiodating, as the technique is sometimes called, is tremendously powerful in determining how fast and when processes happened on the Earth and in the early solar system. Samples of most geological material has been dated: the lunar crustal rocks and basalts returned by the Apollo and Luna missions, all the kinds of meteorites including those from Mars, and tens of thousands of samples from all over the Earth. While the surface of the Moon has been shown to be between 3.5 and 4.6 billion years old for the most part, the Earth's surface is largely younger than 250 Ma (million years old). The oldest rock found on Earth is the Acasta gneiss, from northwestern Canada, which is 3.96 billion years old.

If the oldest rock on Earth is 3.96 billion years old, does that mean that the Earth is 3.96 billion years old? No, because older rocks probably have simply been destroyed by the processes of erosion and plate tectonics, and there is reason to believe that the Earth and Moon formed at nearly the same time. Many meteorites, especially the primitive chondritic meteorites, have ages of 4.56 billion years. Scientists believe that this is the age of the solar system. How, a critical reader should ask, is it known that this is when the solar system formed, and not some later formation event?

(continues)

(continued)

The answer is found by using another set of radioactive elements. These have half-lives so short that virtually all the original parent atoms have decayed into daughters long since. They are called extinct nuclides (nuclide is a synonym for isotope). An important example is ^{129}I (an isotope of iodine), which decays into ^{129}Xe (xenon) with a half-life of only 16 million years. All the ^{129}I that the solar system would ever have was formed when the original solar nebula was formed, just before the planets began to form. If a rock found today contains excess ^{129}Xe, above the solar system average, then it formed very early in solar system time, when ^{129}I was still live. The meteorites that date to 4.56 Ga (billion years) have excess ^{129}Xe, so 4.56 Ga is the age of the beginning of the solar system.

(continued from page 38)

Sun is plotted against the abundance of elements in chondrite meteorites, the plot forms almost a perfect straight line. Other meteorites and especially planetary materials do not have comparable abundances to the Sun. Mars is thought to have formed from chondritic material, similar to what formed the Earth. This makes sense, since they are neighbors in the solar system: The material in this part of the early nebula probably did not differ much between the orbits of Earth and Mars.

The planets must have grown from collisions of matter in the early solar system nebula. Chunks of material collided and stayed together, eventually gathering enough mass to create significant gravity. Growing masses are often referred to as planetesimals, bodies that are too small and evolving too fast to be considered planets yet. These planetesimals continue to collide with each other and grow into a larger body, sweeping up the smaller matter available in the orbit of this growing planet. The early planetesimals were probably irregular; see the sidebar "Accretion and Heating: Why Are Some Solar System Objects Round and Others Irregular?" on page 45. The final planet Mars is round. When and how did that transformation occur?

Volcanic rocks melt from a planet's mantle and so are a direct link to the composition of the planet's interior. Almost all the Martian meteorites are volcanic rocks, so their compositions are the greatest clues to the composition of the interior of Mars. Volcanic rocks on Earth have similar compositions and are made of similar minerals to those delivered from Mars as Martian meteorites, with a few important exceptions. The compositions of the rocks from Mars indicate that they melted from a source that was richer in iron and poorer in aluminum than those from the Earth. A number of estimates for the mantle compositions on Earth and Mars are shown in the table below. This table lists the five major element oxides that make up more than 98 percent of the silicate mantles of the planets.

The first two Martian mantle estimates are about 25 percent lower in alumina (Al_2O_3) than the Earth mantle estimates, and they are also more than twice as rich in iron oxide (FeO). The scientific community today generally agrees that the Martian mantle seems to be depleted in alumina and enriched in iron compared to the Earth's, although the third Martian mantle composition listed indicates an even higher alumina content than the Earth's. The question remains whether Mars

BULK COMPOSITIONS OF EARTH AND MARS					
Source of the model	**SiO_2**	**Al_2O_3**	**FeO**	**MgO**	**CaO**
Earth					
Cl chondritic mantle	49.52	3.56	7.14	35.68	2.82
Hart & Zindler	45.96	4.06	7.54	37.78	3.21
Ringwood	44.76	4.46	8.43	37.23	3.60
McDonough & Sun	45.00	4.45	8.05	37.80	3.55
Allegre et al.	46.12	4.09	7.49	37.77	3.23
Mars					
Bertka and Fei	43.90	3.15	18.80	31.66	2.50
Dreibus and Wanke	45.32	3.08	18.27	30.83	2.50
Morgan and Anders	42.06	6.49	15.98	30.17	5.29

Note: Numbers are expressed in mass percent (mass %).

started with the same composition as the Earth and developed a different mantle composition through different processes of differentiation, or whether the planet began with a different composition.

Because Mars formed in the vicinity of the Earth and Moon, it is unlikely to have formed from material with very different relative elemental abundances. The elements in the solar nebula were likely to be smoothly spread through the nebula, with the heaviest closest to the Sun and the lighter elements farther out. This line of reasoning prompts some scientists to look for processes in early planetary formation that can, in the case of Mars, place more iron and less aluminum in the mantle relative to the core.

A slightly different process of core formation can easily explain the difference in iron content. Early in planetary formation iron and nickel sink through the influence of gravity deep into the planet's interior and form the core. The silicate minerals left behind are less dense and so float on top of the iron, forming the mantle. If Mars had more oxygen or more water present as it formed, more of its iron would stay in the silicate mantle as iron oxide, and less would be reduced to iron metal and sink into the core. If there was less oxygen available on the Earth, then its iron was more likely to stay bound up in silicate mantle materials such as olivine and pyroxene and less likely to form metal iron and sink into the core. If indeed the Earth was struck by a planet-sized asteroid late in its accretionary process, leading to the formation of the Moon and Earth system, then the material making up the Moon and Earth has been heated to very high temperatures and the oxygen and volatiles burned off. Without the oxygen and water, more of the iron in the Earth would sink to the core as metal instead of bonding with oxygen and remaining in the mantle.

The problem of depleted alumina in the Martian mantle is harder to explain. Calling on a bulk composition for Mars that is significantly different from the Earth's challenges current concepts of planetary accretion in the early solar system; adjacent planets should have similar compositions. Still, Mars may

(continues on page 48)

ACCRETION AND HEATING: WHY ARE SOME SOLAR SYSTEM OBJECTS ROUND AND OTHERS IRREGULAR?

There are three main characteristics of a body that determine whether it will become round.

The first is its *viscosity,* that is, its ability to flow. Fluid bodies can be round because of surface tension, no matter their size; self-gravitation does not play a role. The force bonding together the molecules on the outside of a fluid drop pull the surface into the smallest possible area, which is a sphere. This is also the case with gaseous planets, like Uranus. Solid material, like rock, can flow slowly if it is hot, so heat is an important aspect of viscosity. When planets are formed, it is thought that they start as agglomerations of small bodies, and that more and more small bodies collide or are attracted gravitationally, making the main body larger and larger. The heat contributed by colliding planetesimals significantly helps along the transformation of the original pile of rubble into a spherical planet: The loss of their kinetic energy (more on this at the end of this sidebar) acts to heat up the main body. The hotter the main body, the easier it is for the material to flow into a sphere in response to its growing gravitational field.

The second main characteristic is density. Solid round bodies obtain their shape from gravity, which acts equally in all directions and therefore works to make a body a sphere. The same volume of a very dense material will create a stronger gravitational field than a less dense material, and the stronger the gravity of the object, the more likely it is to pull itself into a sphere.

The third characteristic is mass, which is really another aspect of density. If the object is made of low-density material, there just has to be a lot more of it to make the gravitational field required to make it round.

Bodies that are too small to heat up enough to allow any flow, or to have a large enough internal gravitational field, may retain irregular outlines. Their shapes are determined by mechanical strength and response to outside forces such as meteorite impacts, rather than by their own self-gravity. In general the largest asteroids, including all 100 or so that have diameters greater than 60 miles (100 km), and the larger moons, are round from self-gravity. Most asteroids and moons with diameters larger than six miles (10 km) are round, but not all of them, depending on their composition and the manner of their creation.

(continues)

(continued)

There is another stage of planetary evolution after attainment of a spherical shape: internal differentiation. All asteroids and the terrestrial planets probably started out made of primitive materials, such as the class of asteroids and meteorites called CI or enstatite chondrites. The planets and some of the larger asteroids then became compositionally stratified in their interiors, a process called differentiation. In a *differentiated body*, heavy metals, mainly iron with some nickel and other minor impurities in the case of terrestrial planets, and rocky and icy material in the case of the gaseous planets, have sunk to the middle of the body, forming a core. Terrestrial planets are therefore made up, in a rough sense, of concentric shells of materials with different compositions. The outermost shell is a crust, made mainly of material that has melted from the interior and risen buoyantly up to the surface. The mantle is made of silicate minerals, and the core is mainly of iron. The gas giant outer planets are similarly made of shells of material, though they are gaseous materials on the outside and rocky or icy in the interior. Planets with systematic shells like these are called differentiated planets. Their concentric spherical layers differ in terms of composition, heat, density, and even motion, and planets that are differentiated are more or less spherical. All the planets in the solar system seem to be thoroughly differentiated internally, with the possible exception of Pluto and Charon. What data there is for these two bodies indicates that they may not be fully differentiated.

Some bodies in the solar system, though, are not differentiated; the material they are made of is still in a more primitive state, and the body may not be spherical. Undifferentiated bodies in the asteroid belt have their metal component still mixed through their silicate portions; it has not separated and flowed into the interior to form a core.

Among asteroids, the sizes of bodies that differentiated vary widely. Iron meteorites, thought to be the differentiated cores of rocky bodies that have since been shattered, consist of crystals that grow to different sizes directly depending upon their cooling rate, which in turn depends upon the size of the body that is cooling. Crystal sizes in iron meteorites indicate parent bodies from six to 30 miles (10 to 50 km) or more in diameter. Vesta, an asteroid with a basaltic crust and a diameter of 326 miles (525 km), seems to be the largest surviving differentiated body in the asteroid belt. Though the asteroid Ceres, an unevenly-shaped asteroid approximately 577 by 596 miles (930 by 960 km), is much larger than Vesta, it seems from spectroscopic analyses to be largely undifferentiated. It is thought that the higher percentages of volatiles available at the distance of Ceres's orbit may have helped cool the asteroid faster and prevented the buildup of heat required for differentiation. It is also believed that Ceres and Vesta are among the last surviving "protoplanets," and that almost all asteroids of smaller size are the shattered remains of larger bodies.

Where does the heat for differentiation come from? The larger asteroids generated enough internal heat from radioactive decay to melt (at least partially) and differentiate (for more on radioactive decay, see the sidebar called "Elements and Isotopes" on page 27). Generally bodies larger than about 300 miles (500 km) in diameter are needed in order to be insulated enough to trap the heat from radioactive decay so that melting can occur. If the body is too small, it cools too fast and no differentiation can take place.

A source for heat to create differentiation, and perhaps the main source, is the heat of accretion. When smaller bodies, often called *planetesimals,* are colliding and sticking together, creating a single larger body (perhaps a planet), they are said to be accreting. Eventually the larger body may even have enough gravity itself to begin altering the paths of passing planetesimals and attracting them to it. In any case, the process of accretion adds tremendous heat to the body, by the transformation of the kinetic energy of the planetesimals into heat in the larger body. To understand kinetic energy, start with momentum, called p, and defined as the product of a body's mass m and its velocity v:

$$p = mv$$

Sir Isaac Newton called momentum "quality of movement." The greater the mass of the object, the greater its momentum is, and likewise, the greater its velocity, the greater its momentum is. A change in momentum creates a force, such as a person feels when something bumps into her. The object that bumps into her experiences a change in momentum because it has suddenly slowed down, and she experiences it as a force. The reason she feels more force when someone tosses a full soda to her than when they toss an empty soda can to her is that the full can has a greater mass, and therefore momentum, than the empty can, and when it hits her it loses all its momentum, transferring to her a greater force.

How does this relate to heating by accretion? Those incoming planetesimals have momentum due to their mass and velocity, and when they crash into the larger body, their momentum is converted into energy, in this case, heat. The energy of the body, created by its mass and velocity, is called its kinetic energy. Kinetic energy is the total

(continues)

(continued)

effect of changing momentum of a body, in this case, as its velocity slows down to zero. Kinetic energy is expressed in terms of mass m and velocity v:

$$K = \frac{1}{2}mv^2.$$

Students of calculus might note that kinetic energy is the integral of momentum with respect to velocity:

$$K = \int mvdv = \frac{1}{2}mv^2.$$

The kinetic energy is converted from mass and velocity into heat energy when it strikes the growing body. This energy, and therefore heat, is considerable, and if accretion occurs fast enough, the larger body can be heated all the way to melting by accretional kinetic energy. If the larger body is melted even partially, it will differentiate.

How is energy transfigured into heat, and how is heat transformed into melting? To transfer energy into heat, the type of material has to be taken into consideration. Heat capacity describes how a material's temperature changes in response to added energy. Some materials go up in temperature easily in response to energy, while others take more energy to get hotter. Silicate minerals have a heat capacity of 245.2 cal/°lb (1,256.1 J/°kg). What this means is that 245.2 calories of energy are required to raise the temperature of one pound of silicate material one degree. Here is a sample calculation. A planetesimal is about to impact a larger body, and the planetesimal is a kilometer in radius. It would weigh roughly 3.7×10^{13} lb (1.7×10^{13} kg), if its density were about 250 lb/ft³ (4,000 kg/m³). If it were traveling at six miles per second (10 km/sec), then its kinetic energy would be

$$K = \frac{1}{2}mv^2 = \left(1.7 \times 10^{13}\,kg\right)\left(10,000\,m/sec\right)^2$$
$$= 8.5 \times 10^{20}\,J = 2 \times 10^{20}\,cal.$$

(continued from page 44)

have formed from material with less alumina than the Earth. Perhaps the giant impactor that struck the Earth, breaking it apart and allowing it to re-form in the Earth-Moon system,

Using the heat capacity, the temperature change created by an impact of this example planetesimal can be calculated:

$$\frac{8.5 \times 10^{20} \, ^\circ kg}{1,256.1 J / \, ^\circ kg} = 6.8 \times 10^{17} \, ^\circ kg = 8.3 \times 10^{17} \, ^\circ lb.$$

The question now becomes, how much mass is going to be heated by the impact? According to this calculation, the example planetesimal creates heat on impact sufficient to heat one pound of material by $8.3 \times 10^{17} \, ^\circ$F (or one kilogram by $6.8 \times 10^{17} \, ^\circ$C), but of course it will actually heat more material by lesser amounts. To calculate how many degrees of heating could be done to a given mass, divide the results of the previous calculation by the mass to be heated.

The impact would, of course, heat a large region of the target body as well as the impactor itself. How widespread is the influence of this impact? How deeply does it heat, and how widely. Of course, the material closest to the impact will receive most of the energy, and the energy input will go down with distance from the impact, until finally the material is completely unheated. What is the pattern of energy dispersal? Energy dispersal is not well understood even by scientists who study impactors.

Here is a simpler question: If all the energy were put into melting the impacted material, how much could it melt? To melt a silicate completely requires that its temperature be raised to about 2,700°F (1,500°C), as a rough estimate, so here is the mass of material that can be completely melted by this example impact:

$$\frac{6.8 \times 10^{17} \, ^\circ kg}{1,500^\circ} = 4.5 \times 10^{14} \, kg = 9.9 \times 10^{14} \, lb.$$

This means that the impactor can melt about 25 times its own mass ($4.5 \times 10^{14} / 1.7 \times 10^{13} = 26$). Of course this is a rough calculation, but it does show how effective accretion can be in heating up a growing body, and how it can therefore help the body to attain a spherical shape and to internally differentiate into different compositional shells.

contained more alumina than the Earth had begun with, thus enriching Earth's mantle with alumina.

An explanation for Mars's depleted mantle may come from a more detailed examination of the process of differentiation in the early planet. Rather than being a result of initial composition or a result of core formation, the lack of alumina in the regions of Mars that melt to create volcanic rocks may result from the way the planet solidified from its original hot state. In the next section, the concept of a magma ocean is considered: The early planet may have been heated enough that all or most of its silicate mantle was molten, and the process of solidifying could have significant controls on how the planet differentiates.

WAS THERE A MARTIAN MAGMA OCEAN?

The idea that an early planet might heat up to the point that it was mostly or completely molten was first supported by the discovery of the white highlands on the Moon. The Moon's face shows dark circles, which are giant impact craters filled with dark basaltic lava, long since solidified, and white areas between, which are high-standing rock made largely of a silica-rich mineral called plagioclase. This kind of plagioclase has such a calcium-rich composition that it is known under its own name, *anorthite* ($CaAl_2Si_2O_8$). Anorthite is not commonly made in terrestrial lavas, and it exists in astonishingly large volumes on the Moon (the bright white parts of the Moon are almost all anorthite).

Imagine heating a primitive rock such as a chondritic meteorite or a piece of the Earth's mantle until it melts just a few percent. If this melt is then allowed to solidify, it will crystallize into the minerals olivine and pyroxene and just a little plagioclase. This plagioclase would have so much sodium in it that it is not classified as anorthite. How, then, to make a lot of anorthite, if it cannot be solidified from a lava?

The melt would have to go through a lot of compositional change to make it silica-, alumina-, and calcium-rich enough (SiO_2, Al_2O_3, and CaO) to crystallize anorthite. Imagine that as the melt began to cool, all the crystals it formed were removed, perhaps by settling to the bottom of the container the melt was in. The remaining melt would have a different composition than the first melt had: It would have the initial composition with the crystals' composition subtracted. The process of forming crystals and removing them from the melt is called fractionation.

The first minerals to crystallize, olivine and pyroxene, are rich in iron and magnesium, and so the melt they have crystallized (fractionated) from is therefore depleted in iron and magnesium and correspondingly enriched in silica, alumina, and calcium. If this process continues over and over, then the evolving dregs of liquid eventually reach a composition rich enough in calcium, aluminum, and silica that it can crystallize anorthite.

The theory that the lunar scientist John A. Wood of the Smithsonian Astrophysical Observatory proposed in 1970 combined the idea of melt fractionation with the idea that the Moon began as a largely or completely molten ball of magma because of the high energy of the impact that is thought to have created the Earth-Moon system. This is the impact idea: Some time in the first 10s of millions of years of our solar system, enough solid material had clumped together ("accreted") to form the early Earth. Another young planet, about the size of Mars, collided with this early Earth. The collision left behind the metallic core and silicate mantle on the Earth and sprayed off into space a huge amount of extra silicate material. This material reaccreted into the Moon and fell into orbit around the Earth. The heat of the impact and the heat of reaccretion are thought to have melted both the Moon and much of the Earth's mantle, forming magma oceans on both bodies.

As the magma ocean on the Moon began to cool, the magma fractionated olivine and pyroxene, which sank to the bottom of the magma ocean. The remaining liquids evolved into more and more silica- and calcium-rich compositions, until at last they began to crystallize anorthite. Anorthite has an unusual property: It is less dense than the liquid from which it crystallizes, much as ice is less dense than water. All the other minerals that would crystallize from a lunar magma ocean are more dense than their coexisting liquid and would sink. In Wood's model, the anorthite floated to the surface of the magma ocean, forming "rock bergs" (just like icebergs on oceans) that eventually consolidated into the anorthitic crust of the Moon, while the remaining magma ocean liquids finished crystallizing beneath.

This elegant theory explains the white anorthite highlands on the Moon better than any other that has been considered. Nature seldom makes rocks that are almost entirely anorthite, and floating in magma seems to be the simplest explanation. The white highlands on the Moon have become the biggest piece of evidence that magma oceans did exist on the early planets.

Other scientists began studying the idea of a magma ocean as it applies to other terrestrial planets. George Wetherill, of the Carnegie Institute of Washington, noted that there are at least three sources of heat for an early planet: the heat of accretion of the planetesimals that join to make the planet, the potential energy released by the formation of the planet's core when dense metal sinks to the planet's center, and the excess heat given off by the greater amount of radioactive elements present in the early solar system. He and other scientists have proposed that either accretion or core formation alone would be enough to produce energy sufficient to melt the entire silicate portion of the planet.

Other scientists, such as Jay Melosh of the University of Arizona, believe that neither accretion nor core formation were fast enough to create a global magma ocean. He suggests that cooling into space was so fast that those heat sources could not keep up with it, and that instead the giant meteors that were certainly more common in the early solar system would produce heat from their impacts sufficient to produce a series of "magma puddles" or possibly a shallow magma ocean. These impacts may have occurred over a long period of time—see the sidebar called "The Late Heavy Bombardment" on page 54.

The timescales of accretion and core formation are critical to their ability to heat the planet; if either process occurred over a long time period, then its heat could be radiated into space fast enough to prevent melting. The most recent models of accretion predict that Mars accreted in less than 10 million years, which implies that the planet heated to about 2,700°F (1,500°C). At this temperature, the whole planet would have been molten.

Radioactive isotopic systems can help determine the timescales of early solar system processes. The short-lived radiogenic element system tungsten-hafnium (see the sidebar "Determining Age from Radioactive Isotopes" on page 39) is particularly helpful in this regard. In this system, hafnium decays to tungsten (^{182}Hf to ^{182}W) with a half-life of 9 million years. Nine million years is such a short time geologically that it cannot be discriminated using the common uranium-lead dating technique. Nine million years is lost in the haze of the extremely early solar system. The hafnium-tungsten system is useful for studying the time interval between planetary accretion and core formation, because tungsten dissolves into liquid iron and nickel and is carried with them into the core during core formation, while hafnium preferentially stays in the silicate mantle.

If core formation occurred after all the hafnium had decayed, then almost all the tungsten would have been carried into the core and little left in the mantle. If core formation occurred before all the ^{182}Hf had decayed, then the mantle would contain more ^{182}W today, since the hafnium would have decayed in the meantime. There is enough excess tungsten in Martian meteorites to show that the Martian core formed within 10 to 15 million years after planetary accretion, before all the hafnium in the mantle decayed. This estimate of the time of core formation lends support to the magma ocean hypothesis by showing that all the heat of core formation would be available in a short period of time to heat the planet.

A second radiogenic element system, strontium-neodymium, is contained entirely by silicate minerals: Neither strontium nor neodymium dissolve into the iron-nickel core materials. By measuring the isotopes of these elements, it is possible to make a measurement of when the silicate crust of the planet formed, and the answer is again about 20 million years after the solar system formed. Thus the core and crust of Mars formed at about the same time, very early in the life of the solar system. This is an important constraint for scientists making hypotheses about the formation

(continues on page 56)

THE LATE HEAVY BOMBARDMENT

There was a period of time early in solar system development when all the celestial bodies in the inner solar system were repeatedly impacted by large bolides. This high-activity period might be anticipated by thinking about how the planets formed, accreting from smaller bodies into larger and larger bodies, and so it may seem intuitive that there would be a time even after most of the planets formed when there was still enough material left over in the early solar system to continue bombarding and cratering the early planets.

Beyond this theory, though, there is visible evidence on Mercury, the Moon, and Mars in the form of ancient surfaces that are far more heavily cratered than any fresher surface on the planet (Venus, on the other hand, has been resurfaced by volcanic activity, and plate tectonics and surface weathering have wiped out all record of early impacts on Earth). The giant basins on the Moon, filled with dark basalt and visible to the eye from Earth, are left over from that early period of heavy impacts, called the Late Heavy Bombardment.

Dating rocks from the Moon using radioactive isotopes and carefully determining the age relationships of different craters' ejecta blankets indicate that the lunar Late Heavy Bombardment lasted until about 3.8 billion years ago. Some scientists believe that the Late Heavy Bombardment was a specific period of very heavy impact activity that lasted from about 4.2 to 3.8 billion years ago, after a pause in bombardment following initial planetary formation at about 4.56 billion years ago, while other scientists believe that the Late Heavy Bombardment was the tail end of a continuously decreasing rate of bombardment that began at the beginning of the solar system.

In this continual bombardment model, the last giant impacts from 4.2 to 3.8 billion years ago simply erased the evidence of all the earlier bombardment. If, alternatively, the Late Heavy Bombardment was a discrete event, then some reason for the sudden invasion of the inner solar system by giant bolides must be discovered. Were they bodies perturbed from the outer solar system by the giant planets there? If they came from the outer solar system, then more of the material was likely to be water-rich cometary material. If as much as 25 percent of the Late Heavy Bombardment was cometary material, it would have contributed enough water to the Earth to create its oceans. If this model is correct for placing water on the Earth, then a further quandary must be solved: Why didn't Venus receive as much water, or if it did, where did the water go?

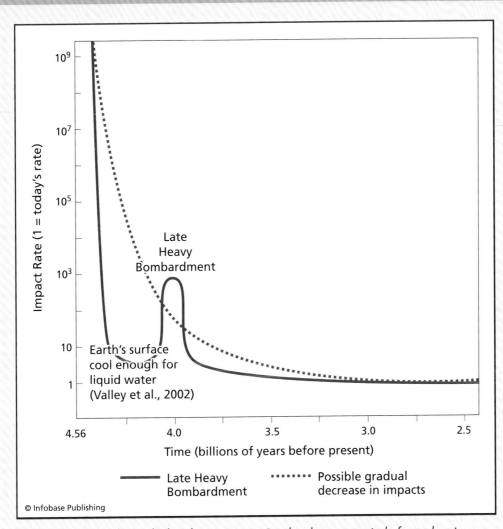

Questions remain about whether the Late Heavy Bombardment, a period of cataclysmic impacts still shown in the giant basalt-filled basins on the Moon, was a discrete event or simply the tail end of clearing out the inner solar system of rubble left over from planet-building. This graph of impact rate v. time demonstrates the two hypotheses.

(continued from page 53)

of Mars: Their theories must allow for early, simultaneous formation of the crust and core.

The author and her colleagues at Brown University, Marc Parmentier and Sarah Zaranek, are studying in detail the process of solidification of a magma ocean. Since all the terrestrial planets probably experienced a magma ocean of some size early in their formation, theories of the processes of solidification may make important predictions about later planetary development. A cooling, crystallizing magma ocean would solidify from the bottom up because of the influence of pressure. At high pressures magma can crystallize at higher temperatures than it can at low surface pressures. Pressure acts to force the atoms together into denser forms, and crystals are almost always denser than the liquid they crystallize from. A magma ocean would be convecting rapidly and keeping its temperature relatively constant from top to bottom, and so it will crystallize first at the high pressures at its bottom.

As crystals form at the bottom of the magma ocean and build up layers of solid, the remaining liquid in the magma ocean is being fractionated and thus is constantly changing. The results of high-pressure experiments in many laboratories around the world, notably Yingwei Fei's laboratory at the Carnegie Institute of Washington, can be used to predict which minerals will form from the magma ocean at a given pressure, temperature, and liquid composition. If the entire silicate mantle of Mars formed the magma ocean, the ocean would have been about 1,240 miles (2,000 km) deep. At its bottom the pressure would be about 24 million atmospheres (24 GPa) (for more, see the sidebar "What Is Pressure?" on page 60). The first minerals to crystallize would have been majorite, an alumina-poor relative of *garnet,* and magnesiowüstite, which is simply a mixture of iron and magnesium oxide. These minerals are only stable at high pressures and though they can be created in the laboratory in special high-pressure furnaces, they have only been found in natural samples as tiny inclusions in diamonds.

At a pressure of about 14 billion atmospheres (14 GPa), which is reached at a depth of about 680 miles (1,100 km) in the Martian magma ocean, majorite and magnesiowüstite would no longer be stable, and the crystallizing minerals would change to garnet, olivine, and pyroxene. Here an unusual thing happens: At these depths olivine and pyroxene are less dense than the silicate liquid and will float, though at lower pressures they are more dense than their coexisting liquid and will sink. As the magma ocean continues to crystallize, then, the garnet will sink while the olivine and pyroxene remain floating in the liquid. Eventually as fractionation continues the olivine and pyroxene will also sink, and in the end, the magma ocean will have fully solidified. The critical consideration at this point is the density profile of the resulting solid Martian cumulate mantle (the solids produced by fractionating a liquid are called cumulates). If higher density materials lie on top of lower density materials, the higher density materials will tend to flow down until the cumulates are ordered from highest density at the bottom to lowest density at the top. (For more, see the sidebar "Rheology, or How Solids Can Flow" on page 79.) Two factors control the density of the cumulate pile. The first is the minerals that make it up, since the differing crystal structure of minerals will produce different densities. The second is the evolution of the liquid magma composition as fractionation proceeds. All the minerals in these magma ocean models contain both magnesium and iron. All the minerals also prefer to incorporate magnesium into their crystal structures rather than iron, and so fractionation preferentially enriches iron in the liquid. As fractionation proceeds, the minerals are forced to incorporate more iron, because the magma is more and more depleted in magnesium. Iron is more than twice as dense as magnesium. The later minerals to crystallize, therefore, are denser than the early minerals, because they contain more iron.

The author's research group has calculated the effects of mineral type and composition on the density of the cumulate stack resulting from Martian magma ocean crystallization and can predict how the cumulates will flow to reach a gravitationally stable order with the densest material at the bottom. The densest cumulates, as it turns out, are those that crystallize last near the surface. These cumulates sink to the bottom of the mantle in great slow-moving solid drips. The next densest material is a part of the original lowermost cumulates, followed by the layer of garnet that is produced while olivine and pyroxene float. This layer of garnet will sink slightly to lie at the top of the densest material. The overturn of cumulates into a stable stratification (densest at the bottom and lightest at the top) will be complete within a few million years of solidification.

Here is a possible answer to the apparent depletion of alumina in Mars's mantle: Garnet, of all the minerals that will crystallize from the magma ocean, contains the most alumina. When it sinks to near the bottom of the cumulate pile, it carries with it most of the alumina in the magma ocean, leaving the rest of the mantle depleted. The later mantle melting that produced the volcanic rocks on Mars's surface is most likely to have come from shallower depths, from which it would carry a record of an alumina-depleted source. In bulk, though, the mantle of Mars would have a similar alumina budget to the mantle of Earth.

Unlike the Moon, Mars does not seem to have any white highlands made of anorthosite plagioclase that floated to the top of its magma ocean. For some time that was taken as evidence that Mars never had a magma ocean. Plagioclase, though, only begins to crystallize once the interior of the planet has solidified up to the point that there is only a shallow magma pond left, about 130 miles (80 km) deep. Though 130 miles may seem a very deep magma pond indeed, the original Martian magma ocean may have been as deep as 3,200 miles (2,000 km). By the time solidification had proceeded to 94 percent completion, when the pond would be only 130 miles (209.2 km) deep, the remaining magma ocean was probably thickly filled with crystals and the plagioclase was unable to float to the top. Scientists think that planets the size of Mars and larger would never form plagioclase flotation crusts, and only relatively small bodies like the Moon can do so.

THE CORE OF MARS

On Earth the radius of the core is known very exactly from the study of how earthquake waves move through the planet. For Mars, there is no such data. Two seismometers were flown to Mars on the Viking mission, though one failed to deploy on the planet. The other seismometer detected one Mars quake, not enough to learn anything about the Martian interior. The core of Mars, therefore, has to be modeled from theory and remote sensing data. Mars's core is thought to be a mixture of iron and iron sulfide. To match the planet's density and size, a certain amount of iron has to be present in the core. Sulfur makes the core material less dense and thus allows a larger core while still matching the density and size requirements. Because of this uncertainty, the radius of Mars's core cannot be constrained very tightly. It has been estimated to be between about 1,100 and 1,370 miles (1,800 and 2,200 km) by Yingwei Fei at the Carnegie Institute. He made the estimates by using a parameter called the moment of inertia factor of the planet. The moment of inertia of a planet is a measure of how much force is required to increase the spin of the planet. (In more technical physics terms, the angular acceleration of an object is proportional to the torque acting on the object, and the proportionality constant is called the moment of inertia of the object.) The moment of inertia depends on the mass of the planet and on how this mass is distributed around the planet's center. The farther the bulk of the mass is from the center of the planet, the greater the moment of inertia. In other words, if all the mass is at the outside, it takes more force to spin the planet than if all the mass is at the center. This is similar to an example of two wheels with the same mass: one is a solid plate, and the other is a bicycle wheel, with almost all the mass at the rim. The bicycle wheel has the greater moment of inertia and takes more force to create the same angular acceleration. The units of the moment of inertia are units of mass times distance squared, for example $lb \times ft^2$ or $kg \times m^2$.

By definition, the moment of inertia I is defined as the sum of mr^2 for every piece of mass m of the object, where r is the

(continues on page 62)

WHAT IS PRESSURE?

The simple definition of pressure (p) is that it is force (F) per area (a):

$$p = \frac{F}{a}.$$

Atmospheric pressure is the most familiar kind of pressure and will be discussed below. Pressure, though, is something felt and witnessed all the time, whenever there is a force being exerted on something. For example, the pressure that a woman's high heel exerts on the foot of a person she stands on is a force (her body being pulled down by Earth's gravity) over an area (the area of the bottom of her heel). The pressure exerted by her heel can be estimated by calculating the force she is exerting with her body in Earth's gravity (which is her weight, here guessed at 130 pounds, or 59 kg, times Earth's gravitational acceleration, 32 ft/sec², or 9.8 m/sec²) and dividing by the area of the bottom of the high heel (here estimated as one square centimeter):

$$p = \frac{(59kg)(9.8\ m/sec^2)}{(0.01^2 m^2)} = 5,782,000\ kg/m/sec^2.$$

The resulting unit, kg/ms², is the same as N/m and is also known as the pascal (Pa), the standard unit of pressure (see appendix 1, "Units and Measurements," to understand more). Although here pressure is calculated in terms of pascals, many scientists refer to pressure in terms of a unit called the atmosphere. This is a sensible unit because one atmosphere is approximately the pressure felt from Earth's atmosphere at sea level, though of course weather patterns cause continuous fluctuation. (This fluctuation is why weather forecasters say "the barometer is falling" or "the barometer is rising": The measurement of air pressure in that particular place is changing in response to moving masses of air, and these changes help indicate the weather that is to come.) There are about 100,000 pascals in an atmosphere, so the pressure of the woman's high heel is about the same as 57.8 times atmospheric pressure.

What is atmospheric pressure, and what causes it? Atmospheric pressure is the force the atmosphere exerts by being pulled down toward the planet by the planet's gravity, per unit area. As creatures of the Earth's surface, human beings do not notice the pressure of the atmosphere until it changes; for example, when a person's ears pop during a plane ride because the atmospheric pressure lessens with altitude. The atmosphere is thickest (densest) at the planet's surface and gradually becomes thinner (less and less dense) with height above the planet's surface. There is no clear break between the atmosphere and space: the atmosphere just gets thinner and thinner and less and less detectable. Therefore, atmospheric pressure is greatest at the planet's surface and becomes less and less as the height above the planet increases. When the decreasing density of the atmosphere and gravity are taken into consideration, it turns out that atmospheric pressure decreases exponentially with altitude according to the following equation:

$$p(z) = p_o e^{-\alpha z},$$

where $p(z)$ is the atmospheric pressure at some height above the surface z, p_o is the pressure at the surface of the planet, and α is a number that is constant for each planet, and is calculated as follows:

$$\alpha = \frac{g\rho_0}{p_0},$$

where g is the gravitational acceleration of that planet, and ρ_o is the density of the atmosphere at the planet's surface.

Just as pressure diminishes in the atmosphere from the surface of a planet up into space, pressure inside the planet increases with depth. Pressure inside a planet can be approximated simply as the product of the weight of the column of solid material above the point in question and the gravitational acceleration of the planet. In other words, the pressure P an observer would feel if he or she were inside the planet is caused by the weight of the material over the observer's head (approximated as ρh, with h the depth you are beneath the surface and ρ the density of the material between the observer and the surface) being pulled toward the center of the planet by its gravity g:

$$P = \rho g h$$

The deeper into the planet, the higher the pressure.

(continued from page 59)

radius for that mass m. In a planet the density changes with radius, and so the moment of inertia needs to be calculated with an integral:

$$I = \int_0^{r_f} \left(\rho(r) \right) r^2 dr$$

where r_o is the center of the planet and r_f is the total radius of the planet, $\rho(r)$ is the change of density with radius in the planet, and r is the radius of the planet, and the variable of integration. To compare moments of inertia among planets, scientists calculate what is called the moment of inertia factor. By dividing the moment of inertia by the total mass of the planet M and the total radius squared R^2, the result is the part of the moment of inertia that is due entirely to radial changes in density in the planet, and it also produces a nondimensional number, because all the units cancel. The equation for the moment of inertia factor, K, is as follows:

$$K = \frac{I}{MR^2}.$$

The issue with calculating the moment of inertia factor for a planet is that, aside from the Earth, there is no direct information on the density gradients inside any planet. There is another equation, the rotation equation, which allows the calculation of moment of inertia factor by using parameters that can be measured. This equation gives a relationship between T, the rotation period of the planet in seconds; K, the moment of inertia factor of the planet in kg/°m³; M, the mass of the planet in kg; G, the gravitational constant in kg; R, the planet's polar radius in m; D, the density for the large body in kg/m³; a, the planet's semimajor axis in m; i, the orbital inclination of the planet in degrees; m, the total mass of all satellites that orbit the large body in kg; d, the mean density for the total satellites that orbit large body in kg/m³; and r, the mean polar radius of all satellites that orbit the large body in meters:

$$T^2 = K \left(\frac{4\pi^2 D^3}{G(M+m)\cos^2 i} \right) \left(\left(\frac{m}{M} \right) \left(\frac{r}{R} \right) + \left(\frac{M}{m} \right) \left(\frac{D}{d} \right) \right) (4\pi a r K).$$

By getting more and more accurate measures of the moment of inertia factor of Mars from these external measurements, scientists can test their models for the interior of Mars. By integrating their modeled density structures, they can see whether the model creates a moment of inertia factor close to what is actually measured for Mars. On the Earth, the moment of inertia factor can be used to test for core densities, helping constrain the percentage of light elements that have to be mixed into the iron and nickel composition.

MARTIAN TIME LINE OF EVENTS

Because the radius of Mars is about half that of Earth, early Mars probably both heated up and is cooling off faster than the Earth. On Earth the continued release of radiogenic and accretionary heat from the interior causes convection in the mantle and the movement of plates across the surface of the planet. Mars, by comparison, is a one-plate planet: The exterior of Mars is a single, rigid shell. Earlier in Mars's history, it is possible that the planet's interior moved enough and was hot enough to allow the existence of only a thin outer crustal shell and possibly plate tectonics. The Martian meteorites, which are almost all volcanic in origin, show that there has been volcanic activity on Mars for billions of years and probably into the present (the most recent eruption appears to be only about 10 million years old). Unfortunately, scientists know exact dates for almost no Martian events.

On Earth the age of a rock can often be determined exactly by measuring its radioactive isotopes and their daughter products, and thereby knowing how long the radioactive elements have been in the rock, decaying to form their daughters. Before the discovery of radioactivity and its application to determining the age of rocks, all of which happened in the 20th century, geologists spent a couple of centuries working out the relative ages of rocks, that is, which ones were formed first, and in what order the others came. Between the years of 1785 and 1800, James Hutton and William Smith introduced and labored over the idea of geologic time: The rock record describes events that happened over a long time period.

Fossils were the best and easiest way to correlate between rocks that did not touch each other directly. Some species of fossil life can be found in many locations around the world and so form important markers in the geologic record. Relative time was broken into sections divided by changes in the rock record, for example, times when many species apparently went extinct, since their fossils were no longer found in younger rocks. This is why, for example, the extinction of the dinosaurs lies directly on the Cretaceous-Tertiary boundary: The boundary was set to mark their loss. The largest sections of geologic history were further divided into small sections, and so on, from epochs, to eras, to periods. For centuries a debate raged in the scientific community over how much time was represented by these geologic divisions. With the development of radioactive dating methods, those relative time markers could be converted to absolute time; for example, the oldest known rock on Earth is 3.96 billion years old, and the Cretaceous-Tertiary boundary lies at about 66.5 million years ago.

With the possible exception of the Moon, the Earth is the only body for which scientists can develop a detailed relative timescale because there has never been a field geologist on any other planet to do the necessary careful mapping of geologic units. A number of absolute radioisotope dates have been made for lunar rocks returned from the Apollo missions, forming the beginning of an absolute timescale for the Moon. Scientists cannot develop an absolute timescale for Mars because the only rocks available are the Martian meteorites, whose original locations on Mars are unknown. For other bodies there are no absolute dates. Using detailed images of other planets, though, it is possible to work out the relative ages of many of the crustal features. By carefully examining images, researchers can determine "superposition," that is, which rock unit was formed first, and which later came to lie on top of it. Impact craters and canyons are very helpful in determining superposition. Through this sort of meticulous photogeology, scientists have developed relative timescales for other planets, as shown in the figure on page 65.

A rough set of geological epochs has been built up for Mars and is shown in the figure along with those for the Moon and Earth. From oldest to youngest, they are called the Noachian,

Hesperian, and Amazonian, named after certain rock units. Though photogeology and theory suggest where in Martian history events occurred, only Martian meteorites have

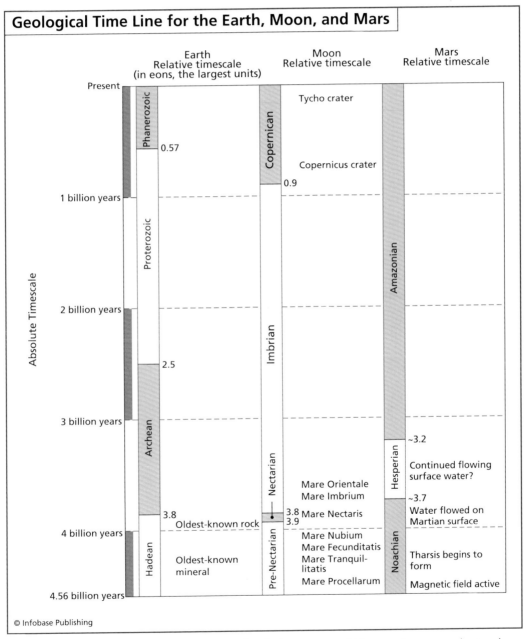

Geological Time Line for the Earth, Moon, and Mars

The geologic time line for the Earth compared with the Moon and Mars as they are understood without further samples to date using radioisotopes

radiometric ages to place them more definitively on the time line. All other events are approximate.

THE CRUST OF MARS AND THE HEMISPHERIC DICHOTOMY

The hemispheric dichotomy, as the obvious difference between the crusts of the northern and southern hemispheres is called, consists of both a difference in elevation and a difference in surface geology. The southern hemisphere is about one-half to three miles (one to three km) higher than the northern hemisphere and the south pole is nearly four miles (six km) higher than the north pole. What is really meant by one pole of a planet being higher than the other? The height of a pole is measured as the distance from the planet's center of mass, not center of volume, to the point on the surface that is being considered. On Mars, the south pole is farther from Mars's center of mass than is the north pole. The overall strong slope of the planet's surface toward its north pole implies that if liquid water once flowed on its surface, water would flow toward and pool in the north.

The southern hemisphere is heavily cratered. The time of heaviest meteorite bombardment was early in the age of the solar system (see the sidebar "The Late Heavy Bombardment" on page 54). The density (number of craters per surface area) and size of craters in Mars's southern hemisphere indicate that its surface is ancient, that it existed during early times when the rate of cratering was higher, and that it has survived since then, collecting additional craters. Its deeply incised channels are another indicator of the great age of the southern hemisphere. Channels take a long time to form, and had to have been cut at a time when Mars's surface was different than it is today, probably during a time when water flowed on the surface. The northern hemisphere, in contrast, is much smoother and is thought to have been resurfaced early in Martian history. The northern hemisphere was long thought to be almost completely smooth and devoid of craters, but now it has been shown that even the flat northern plains hold faint outlines of old craters. The faint outlines prove that the plains are old, as

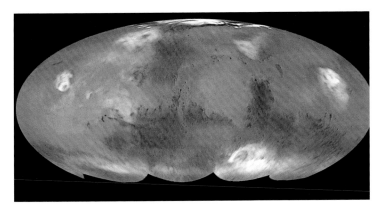

Four hemispheric views have been combined into a full-color global map showing the regions of Mars imaged by the Hubble Space Telescope *during the planet's closest approach to Earth. Latitudes below about 60 degrees south were not viewed by the telescope because the planet's north pole was tilted toward Earth.* (NASA/ Cornell University/University of Colorado)

old as the southern highlands. Mars Orbital Laser Altimeter (MOLA; see the sidebar "Mars Orbital Laser Altimeter" on page 71) data was used to find these craters in 2002. Since the time of highest cratering rates the northern plains have been resurfaced, perhaps by a layer of dust that caps and smooths all the craters.

MOLA mapped the topography of Mars in great detail, greater detail, in fact, than is commonly available for Earth. (Government security agencies may have similarly detailed topography of Earth, but it is classified and unavailable and remains at the level of rumors.) An example of the science team's results is shown in the accompanying map of Mars. The topographic heights of the Tharsis rise and their attendant immense outflow valley, Valles Marineris, can be seen in the middle right of the figure. The large Hellas impact basin is almost perfectly opposite Tharsis on Mars. The smooth northern plains and the cratered, topographically high southern plains are also apparent in this figure.

Originally, when the northern plains looked smooth (there were no photos or altimetry data good enough to detect the buried craters), the plains were thought to be much younger than the southern hemisphere. If the northern plains were not younger, then water movement and sedimentation in oceans, volcanic eruptions, or wind-blown sediments had to have resurfaced them, because of their smoothness. A simpler answer seemed to be that the crust itself was younger and had not existed during the time of heaviest cratering. When the MOLA data was first analyzed, and many ancient crater rings could be detected through the smooth surface veneer of the northern plains, researchers thought the north and south were about the same age. Sediments have draped over the craters, softening their outlines until they were difficult to see in low-resolution images. With further analysis, it is clear that there are still fewer craters in the northern lowlands than there are in the southern highlands, and so the north is a bit younger than the south.

The northern lowlands are now thought to date to the early Noachian, and the southern highlands to be a little older. (For more on the Martian timescale, see the previous section "Martian Time Line of Events" on page 63.) The buried crater Utopia in the northern hemisphere requires that the crust beneath the northern plains be Noachian in age, as Utopia had to have been made by an impactor of such size that almost none were left after the main phase of planetary accretion.

There are a variety of hypotheses on the formation of the differences between the northern and southern hemispheres on Mars. One hypothesis formulated in the mid-1990s suggests that there had been strong enough plate tectonics on Mars in its early days that crustal spreading, such as there is on Earth at mid-ocean ridges, created the lower-lying, smooth northern plains. On Earth, the major ocean basins are created by such spreading centers, where hot mantle material is welling up and melting to some extent, and the melt is rising to the surface and cooling to form new oceanic crust. This crust is thinner, smoother, and denser than continental crust, and so it lies lower than continental crust and water naturally fills these broad, low-lying basins, creating oceans. The Pacific, Atlantic, and Indian Oceans all have long, sinuous spreading centers like seams up their centers. New oceanic crust spreads out to each side. This crust is thinner than almost any continental crust on Earth. Thus, by analogy, it was suggested that the northern lowlands on Mars were simply a huge ocean basin created by seafloor spreading.

A second hypothesis for the northern lowlands is that they were created by one or more giant impacts. This hypothesis is supported by the possible detection of large circular structures that could correspond to impacts of the size required, but it may be disproved by MOLA data that shows that crustal structure is roughly the same everywhere on Mars, that the northern and southern crusts are almost equally cratered, and that the alleged faint and ancient giant impacts have no gravity anomalies associated with them. (Impacts excavate material from the crater region and therefore create differences in the gravity field inside the crater versus outside.) The MOLA data also works against the seafloor spreading theory: Crust produced in that way should be markedly thinner and denser than the rest of the planet's crust.

For a time then, the idea that the northern plains were created by a number of giant impact craters that fortuitously overlapped was no longer the favorite theory. In 2008, however, a scientist at MIT named Jeffrey Andrews-Hanna had a new approach to the impact idea. Rather than a number of lucky direct hits, what if the northern plains were created by one super-giant glancing blow that just knocked off the upper crust? The northern plains, sometimes called the Borealis basin, make up 40 percent of the surface of Mars. The basin—13,600 miles (8,500 km) by 17,000 miles (10,600 km)—is larger than the combined area of Asia, Europe, and Australia. The formation of this giant elliptical basin has been successfully modeled by computer codes that simulate impacts on planets.

Formation of the northern lowlands by impacts is known as exogenic formation, that is, from an external cause. Other hypotheses call on internal processes to create the dichotomy, and these hypotheses are called endogenic. The internal process hypotheses use arguments involving mantle convection. A very viscous material like a planetary mantle convects (circulates) in response to heat from the interior. (For more on these topics, see the sidebar "Rheology, or How Solids Can

Flow" on page 79.) As the planet loses heat to space, shallow layers become cool. The coolest, shallowest layers are less dense than the warmer materials below them, because cooling most materials causes them to contract very slightly and thus to be more dense. These cool, dense layers may then start sinking downward, displacing warmer, less dense material below them. When the cool, sinking material reaches the bottom of the mantle, it may heat by conduction from the hot core. Warm and buoyant again, this material may rise back into the shallow mantle, only to be cooled by surface heat loss and sink again. These hot upwellings and cool downwellings can form what are called cells, where the upwellings and downwellings form a continuous cycle more or less stationary in the mantle, or the patterns of upwellings and downwellings can be more chaotic and quickly changing.

The geophysicists Shijie Zhong at the University of Colorado, Maria Zuber at the Massachusetts Institute of Technology, and E. M. Parmentier at Brown University used computer codes to model how the interior of Mars might convect. They found that convection in the Martian mantle is likely to arrange itself into one planetwide convection cell. This means that the downwelling part of the cell would take up one hemisphere of the planet, perhaps the northern hemisphere, and the upwelling would take up the other hemisphere, in this example, the southern. This pattern of convection is called "mode 1" because it is the first, simplest pattern of convection in a spherical shell. The downwelling would strip cool material from the bottom of the northern lithosphere, thinning and pulling it downward. The upwelling would buoyantly support the southern lithosphere and might melt during decompression and create additional crust. In this way, the crustal dichotomy could be created by mantle convection.

A third theory for the Martian crustal dichotomy involves a similar convection pattern, but at a different point in Martian history and for a different reason. The process of crystallizing the early Martian magma ocean has already been described. When the planet is fully crystalline, the mantle is prone to overturning because the deep layers of the crystalline

(continues on page 74)

MARS ORBITAL LASER ALTIMETER

The Mars Orbital Laser Altimeter (MOLA) is an instrument currently in orbit around Mars on the *Mars Global Surveyor* spacecraft. MOLA can measure the range (distance) to the Martian surface with a precision of 14 inches (37 cm), and a profiling resolution on the Martian surface of about 980 feet (300 m). This resolution means that no greater definition can be made of the surface than a measurement every 980 feet (300 m), though each individual elevation measurement is far more accurate. To determine the distance to the planet with this accuracy, the exact location of the spacecraft must also be known. At worst, the height of the *Mars Global Surveyor*'s orbit is known to an error of less than 16 feet (5 m).

(continues)

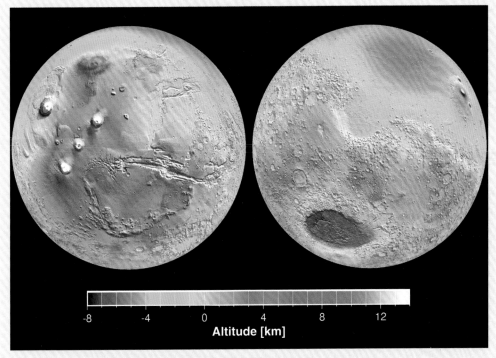

This false-color topographic map of Mars was created using data from the Mars Orbital Laser Altimeter and has an accuracy of less than 15 feet in elevation (five m). The Hellas impact basin is in purple, and the huge Tharsis rise in red and white; note also the smooth northern lowlands. (NASA/JPL)

(continued)

The Mars Global Surveyor mission was launched November 7, 1996, by NASA and the Jet Propulsion Laboratory of the California Institute of Technology, and it reached Mars orbit on September 12, 1997. It was at the time the first successful Mars mission in 20 years. The *Mars Global Surveyor* mapping orbit, which was attained in February 1999 after a year and a half of adjustments, is nearly circular, with an inclination of 92.869 degrees, and an eccentricity of 0.00405. The spacecraft altitude ranges from 234 miles (375 km) to 278 miles (445 km) above the surface. The orbital period is about 117 minutes.

To make measurements, the instrument fires short (eight nanosecond) infrared laser pulses toward Mars at a rate of 10 Hz (see the appendix called "Light, Wavelength, and Radiation") and measures the time for the reflected energy to return to the instrument. By dividing this travel time by two (to obtain just the distance to the surface rather than the length of the whole round trip) and by the velocity of the laser light, the MGS spacecraft measures the range (distance) to the tops of Mars's mountains and the depths of its valleys.

In detail, a telescope focuses the light scattered by terrain or clouds onto a silicon avalanche photodiode detector. A portion of the output laser energy is diverted to the detector and starts a precision clock counter. The detector outputs a voltage proportional to the rate of returning photons that have been backscattered from Mars's surface or atmosphere. When this voltage exceeds a noise threshold, the detector stops the clock, and the time interval is used to calculate the distance to the surface. Four separate channels filter the voltage output in parallel to detect pulses spread out by the roughness of terrain or clouds, increasing the likelihood of detection, and providing some information about surface or cloud characteristics. Range measurements have been used to construct precise topographic maps of Mars that have many applications to studies in geophysics, geology, and atmospheric circulation.

MOLA has not collected altimetry data since June 30, 2001, when the oscillator, the instrument controlling laser firing, ceased operating. At the time of the oscillator anomaly, MOLA had been in space for 1,696 days. The MOLA laser had fired 671 million times, and MOLA had made about 640 million measurements of the Martian surface and atmosphere. The June 30 event was the first anomaly in MOLA's operation since the *Mars Global Surveyor* launch in November 1996. The *Mars Global Surveyor* has collected far more information on Mars than all previous missions combined.

As of 2004, the *Mars Global Surveyor* is still collecting image and spectrometer data in an extended mission status. In February 2004 the *Mars Global Surveyor* was

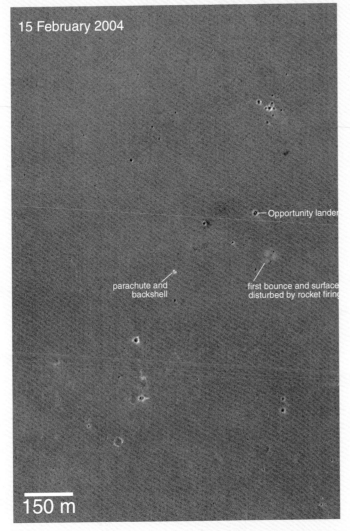

15 February 2004

parachute and
backshell

Opportunity lander

first bounce and surface
disturbed by rocket firing

150 m

On February 15, 2004, the Mars Global Surveyor *spacecraft* flew almost directly over the Mars Exploration Rover Opportunity's landing site and took this image. (NASA/JPL/Malin Space Science Systems)

able to capture a picture of the Mars Exploration Rover *Opportunity* sitting in the small crater where it landed, prior to moving out across the Meridiani plain. The image above shows the location of the lander in its small impact crater, the locations of the parachute and backshell, and the area disturbed by landing rockets and the first bounce. The heat shield impact site was too far east for the camera to view.

(continued from page 70)
mantle are less dense than the upper layers. In this scenario, the mantle overturns (the deep layers rise and the shallow layers sink), but the pattern of the overturn is in mode 1, such that downwelling takes up one hemisphere, and upwelling the other. The majority of the young crust would be created in the south. At the moment, theories for the crustal dichotomy remain in competition, until additional supporting data can be found for one or another.

Both these final theories, involving internal convection, imply that the hemispheric dichotomy consists of more than surface elevation and composition, but that it also reflects fundamental differences in crustal thickness between the hemispheres. The topographic maps of Mars show tremendous heights (the Tharsis volcanic region most predominantly, at about 5.5 miles or nine km height), and low areas (notably the Valles Marinaris, at four miles or seven km depth). As discussed, the warm solid interior of a planet can flow like a liquid over very long times. If there is a high peak of material in the crust resting on the mantle, the lower crust and mantle will eventually flow away, allowing the peak to spread out and to sink. Knowing the height and width of the *relief* on Mars and the approximate age of the high features being studied, the minimum viscosity of the lower crust can be calculated that allows the relief to be maintained as seen in the present. In other words, the lower crust and mantle have to be stiff enough (have high enough viscosity) to have prevented Tharsis from flowing outward and sinking down. By making similar calculations all over the surface of Mars, using MOLA topographic data, the thickness and viscosity of the crust can be calculated.

The crust is thinned under craters, as expected by crater excavation, and thickened under volcanoes, as expected from lava deposits. As discussed above, there are also hemispheric differences: the crust is thicker in the southern hemisphere, with an outside maximum thickness of 50 miles (80 km), and a more average value of 30 to 40 miles (50 to 60 km). In the northern hemisphere, the crust is thought to be about 12 miles (20 km) thinner. Further analysis by the MOLA scientists have shown that the line around the planet where the thick southern crust merges into the thinner northern crust does not occur at the same point that the surface features change from southern to northern. They argue that this indicates that the surface expression of the hemispheric dichotomy is primarily due to surficial, such as oceanic, rather than internal processes, such as convection. This does not invalidate the convection models for creating the differing

The Mars Orbital Laser Altimeter first showed that the apparently smooth northern plains actually consist of a smooth younger surface that lies draped over a heavily cratered surface. The extent of buried craters, here shown in the Utopia region, shows that the northern lowlands have crust of almost the same great age as the southern highlands. (NASA/JPL/ Malin Space Science Systems)

500 m

crustal thicknesses, but it does indicate that, whatever created the differing crustal thicknesses, further surface processes changed its appearance.

The northern hemisphere may be smoother today because it was once a huge ocean; researchers have identified the location of possible shorelines, which may follow the wavy line on the surface between the northern and southern hemispheres. Once shorelines had been identified, a problem quickly became apparent: The shorelines as they are marked today lie across many elevations, wavering up and down by more than a mile (several kilometers) in elevation. By definition, the shoreline of a body of water must lie all at one elevation (such as sea level on Earth) because water will flow to the lowest point and maintain a flat surface. So how could the shoreline wave up and down in elevation?

Taylor Perron, a scientist at the time at the University of California at Berkeley, suggested a fascinating explanation for the wandering shorelines. When planets spin, they bulge at their equators, and the mass of the planet experiences a force to move it into symmetry with the spin axis. Imagine a planet with a very heavy belt around it—if it could move, the belt would migrate to align with the equator of the planet. In fact, the axis of spin of the planet would change, too, to allow this readjustment. This process is called "true" polar wander ("apparent" polar wander, by comparison, is created by the changing magnetic field within the Earth and not by changing the actual spin axis of the planet).

Perron and his colleagues suggested the same thing happened on Mars. The low area, potentially an ocean, migrated up to its current position around the north pole to meet the symmetry requirements of the spinning planet. They migrated the shorelines back and found a case where all the shorelines were at the same elevation before the hypothetical wander! They showed in this way that the basin may actually have been an ocean after all.

Liquid water oceans are also a great place for the formation of life, and so this possibility is an exciting one from many points of view. One of the expectations, if there had been an ocean, is that rock outcroppings made of carbonate minerals would be found. On Earth, carbonates such as calcite ($CaCO_3$) are made in large quantities only in the oceans, and they are produced over a wide variety of temperatures and depths. The *Mars Global Surveyor* spacecraft, as well as other orbiters, has an infrared *spectrometer* that allows the detection of carbonates. (For more, see the sidebar called "Remote Sensing" on page 102.)

In 2003, Timothy Glotch, Joshua Bandfield, and Philip Christensen, scientists at Arizona State University, announced that despite their intensive efforts no large carbonate outcrops were found on Mars. What they have found is just a few per-cent of carbonate material in the surface dust, which shows first that their detector is working well, and second, that there were apparently no early oceans that laid down large deposits of carbonate.

The lack of carbonates is superficially a blow to the idea that there were early oceans on Mars, but instead may indi-cate that the oceans lacked the correct chemistry to deposit carbonate. Alternatively, large carbonate deposits simply may be hidden from the spectrometers in ways not yet thought of. More recent data shows that the Martian surface is highly acidic in many places, at a level that would have prevented the formation of carbonates. There are, however, clearly identified deposits of hematite, an oxidized iron mineral that could only have formed with surface water. One of the Mars Exploration Rovers is exploring one of these hematite regions. Other areas show vast outcrops of sulfate minerals that are likely to have been deposited by a shallow acidic ocean.

In 2008 the *Phoenix* lander was also able to measure the pH (a measure of how acidic the material is) of the soil at its landing site. Much of Mars's surface is rich in sulfur, an indication of acidic conditions, and the large deposits of limestone (calcium carbonate) that were expected as a consequence of ancient oceans have never been found. For these reasons many scien-tists have concluded that Mars's surface is rich in acidic brines as a result of volcanic emissions of sulfur and other chemicals, and that if there ever were limestone deposits they have been dissolved by the acid. *Phoenix*, however, measured very basic (the opposite of acidic) soil at its landing site and confirmed the presence of carbonate. Mars's surface is apparently highly variable in its surface chemistry and reflects many different water interactions.

Simultaneously, the *Mars Express* orbiter was measuring surface compositions using its OMEGA remote sensing equipment. The orbiter was able to detect calcium carbonate in the northern plains near *Phoenix,* showing that the nonacidic regions are widespread and in great contrast to the acidic equatorial regions. These carbonate minerals, along with other minerals detected by OMEGA, all require water for their formation. Finally, in 2009 *Mars Reconnaissance Orbiter* detected strong carbonate signals using its CRISM instrument. These carbonates, however, are near Mars's equator and in the southern highlands, rather than in the remote northern plains. They lie north of Hellas crater, in a region called Nili Fossae. The carbonate outcrops are near other rocks that contain olivine and other silicate minerals and are thought to be magnesium carbonates, rather than the more familiar calcium carbonates. This has led the mission scientists to think that the carbonates there formed from the interaction of groundwater with existing rocks. Their find is further evidence that the acidic Martian surface environment did not extend everywhere on the surface, or these rocks would have been destroyed.

Some scientists are drawing parallels between the sulfur cycle on Mars and the carbon cycle on Earth. Mars's mantle and crust may both contain about twice as much sulfur as their corresponding parts of Earth. On Earth as on Mars, the planetary mantle melts and erupts through volcanoes, which give off sulfur-rich gases along with magma, probably by far the most significant way that sulfur is added to the planetary surface. On the Earth the action of plate tectonics replaces some, when oceanic plates sink in subduction zones and are recycled back into the mantle. On Mars there is no plate tectonics and so whatever is erupted into the atmosphere remains on the planetary surface.

Volcanism on Mars was certainly more common in the distant past than it is today. The planet is paved with lava, but there is little that is fresh, and mankind has not observed a volcano erupting. Mars is not volcanically dead, but it has slowed

(continues on page 80)

RHEOLOGY, OR HOW SOLIDS CAN FLOW

Rheology is the study of how materials deform, and the word is also used to describe the behavior of a specific material, as in "the rheology of ice on Ganymede." Both ice and rock, though they are solids, behave like liquids over long periods of time when they are warm or under pressure. They can both flow without melting, following the same laws of motion that govern fluid flow of liquids or gases, though the timescale is much longer. The key to solid flow is viscosity, the material's resistance to flowing.

Water has a very low viscosity: It takes no time at all to flow under the pull of gravity, as it does in sinks and streams and so on. Air has lower viscosity still. The viscosities of honey and molasses are higher. The higher the viscosity, the slower the flow. Obviously, the viscosities of ice and rock are much higher than those of water and molasses, and so it takes these materials far longer to flow. The viscosity of water at room temperature is about 0.001 Pas (pascal seconds), and the viscosity of honey is about 40 Pas. By comparison, the viscosity of window glass at room temperature is about 10^{27} Pas, the viscosity of warm rocks in the Earth's upper mantle is about 10^{19} Pas.

The viscosity of fluids can be measured easily in a laboratory. The liquid being measured is put in a container, and a plate is placed on its surface. The liquid sticks to the bottom of the plate, and when the plate is moved, the liquid is sheared (pulled to the side). Viscosity is literally the relationship between shear stress σ and the rate of deformation ε. Shear stress is pressure in the plane of a surface of the material, like pulling a spatula across the top brownie batter.

$$\eta = \frac{\sigma}{\varepsilon}.$$

The higher the shear stress needed to cause the liquid to deform (flow), the higher the viscosity of the liquid.

The viscosity of different materials changes according to temperature, pressure, and sometimes shear stress. The viscosity of water is lowered by temperature and raised by pressure, but shear stress does not affect it. Honey has a similar viscosity relation with temperature: The hotter the honey, the lower its viscosity. Honey is 200 times less viscous at 160°F (70°C) than it is at 57°F (14°C). For glass, imagine its behavior at the glasshouse. Glass is technically a liquid even at room temperature, because its molecules are not organized into crystals. The flowing glass the glassblower works with is

(continues)

(continued)

simply the result of high temperatures creating low viscosity. In rock-forming minerals, temperature drastically lowers viscosity, pressure raises it moderately, and shear stress lowers it, as shown in the accompanying figure.

Latex house paint is a good example of a material with shear-stress dependent viscosity. When painting it on with the brush, the brush applies shear stress to the paint,

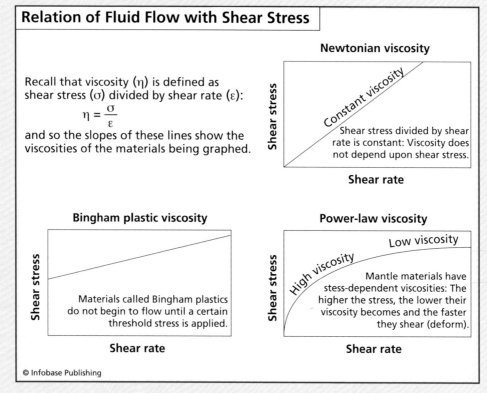

Relation of Fluid Flow with Shear Stress

Recall that viscosity (η) is defined as shear stress (σ) divided by shear rate (ε):

$$\eta = \frac{\sigma}{\varepsilon}$$

and so the slopes of these lines show the viscosities of the materials being graphed.

Newtonian viscosity

Shear stress / Shear rate

Constant viscosity

Shear stress divided by shear rate is constant: Viscosity does not depend upon shear stress.

Bingham plastic viscosity

Shear stress / Shear rate

Materials called Bingham plastics do not begin to flow until a certain threshold stress is applied.

Power-law viscosity

Shear stress / Shear rate

Low viscosity

High viscosity

Mantle materials have stess-dependent viscosities: The higher the stress, the lower their viscosity becomes and the faster they shear (deform).

© Infobase Publishing

These graphs show the relationship of fluid flow to shear stress for different types of materials, showing how viscosity can change in the material with increased shear stress.

(continued from page 78)

way down from its original activity. When sulfur combines with water it makes sulfuric acid, and all over the Martian surface the lava shows evidence of alteration from sulfuric acid. Sulfuric acid effects continued after volcanism waned,

and its viscosity goes down. This allows the paint to be brushed on evenly. As soon as the shear stress is removed, the paint becomes more viscous and resists dripping. This is a material property that the paint companies purposefully give the paint to make it perform better. Materials that flow more easily when under shear stress but then return to a high viscosity when undisturbed are called thixotropic. Some strange materials, called dilatent materials, actually obtain higher viscosity when placed under shear stress. The most common example of a dilatent material is a mixture of cornstarch and water. This mixture can be poured like a fluid and will flow slowly when left alone, but when pressed it immediately becomes hard, stops flowing, and cracks in a brittle manner. The viscosities of other materials do not change with stress: Their shear rate (flow rate) increases exactly with shear stress, maintaining a constant viscosity.

Temperature is by far the most important control on viscosity. Inside the Earth's upper mantle, where temperatures vary from about 2,000°F (1,100°C) to 2,500°F (1,400°C), the solid rocks are as much as 10 or 20 orders of magnitude less viscous than they are at room temperature. They are still solid, crystalline materials, but given enough time, they can flow like a thick liquid. The mantle flows for a number of reasons. Heating in the planet's interior makes warmer pieces of mantle move upward buoyantly, and parts that have cooled near the surface are denser and sink. The plates are also moving over the surface of the planet, dragging the upper mantle with them (this exerts shear stress on the upper mantle). The mantle flows in response to these and other forces at the rate of about one to four inches per year (two to 10 cm per year).

Rocks on the planet's surface are much too cold to flow. If placed under pressure, cold, brittle surface rocks will fracture, not flow. Ice and hot rocks can flow because of their viscosities. Fluids flow by molecules sliding past each other, but the flow of solids is more complicated. The individual mineral grains in the mantle may flow by "dislocation creep," in which flaws in the crystals migrate across the crystals and effectively allow the crystal to deform or move slightly. This and other flow mechanisms for solids are called plastic deformations, since the crystals neither return to their original shape nor break.

as can be seen in the Gusev crater by *Spirit* rover observations. In Gusev crater, rocks show abrasion by wind and sand, meaning that water was not longer common on the surface, and at the same time they show weathering by sulfur. So long after volcanism ceased and surface water froze, impacts and

A pattern of convection in Mars's mantle called mode-1 convection may be responsible for the formation of the crustal dichotomy early in Mars's history.

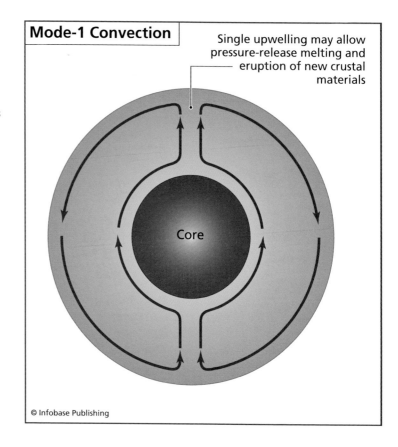

Mode-1 Convection

Single upwelling may allow pressure-release melting and eruption of new crustal materials

Core

© Infobase Publishing

brines and wind moved sulfur around and allowed it to continue reacting with surface rocks.

The emerging image of Mars's history is that there was once flowing water on its surface, followed by a long period of acidic, briny surface conditions, and now finally to the cold desert it currently is. Perhaps, thanks to the new measurements from *Phoenix* and *Mars Express,* there may have been a carbonate-rich period before the acids became common. This stark, global change in chemistry is not seen in Earth's history with the same dramatic relief, perhaps in part because of the tempering influence of life, water, and plate tectonics.

GRAVITY MEASUREMENTS ON MARS

Every planet has different average radii on its equator and its poles; the equatorial radius is larger than the polar radius, because the planet is slightly flattened by spinning. If a person stands on the equator, then, they are slightly farther from the center of the planet than if they were standing on a pole. To a lesser extent, the distance the person is from the center of the planet changes according to whether they are standing on a mountain or in a valley. Being at a different distance from the center of the planet means the person has a different amount of mass between them and the center of the planet. What effect does all the mass have? It pulls with its gravity. (For more information on gravity, see the sidebar called "What Makes Gravity?" on page 84.) If the person stands on the equator, where the radius of the planet is larger and the amount of mass beneath them is relatively larger, the pull of gravity is actually stronger than it is at the poles. Gravity is actually not a perfect constant on any planet: variations in radius, topography, and the density of the material underneath make the gravity vary slightly over the surface. This is why planetary gravitational accelerations are generally given as an average value on the planet's equator.

The *Mars Global Surveyor* made gravity measurements of Mars in a clever way: Whenever the *Global Surveyor* was in a position so that a part of Mars's surface lay directly between it and the Earth, it would send radio signals to Earth. How the radio signals were deflected as they passed Mars is dependent upon the gravity field of Mars, and in this way, the field could be calculated. The average acceleration of gravity at Mars's equator is 3.69 m/sec^2, but since the planet is not a uniform sphere, and because its internal density varies from place to place, so does its gravity field. A gravity field's deviations from average are measured in a unit called the galileos (Gal), or in this case, milliGals (0.001 of a galileo). One galileo is about 0.03 ft/sec^2 (0.01 m/sec^2). A milliGal is 10^{-5} m/sec^2. While changes in gravity due to topography and near-surface density changes are on the order of tens to thousands of mGals, the variation in gravity from pole to equator, due to changing planetary radius, can be on the order of thousands of mGals or more.

The *Mars Global Surveyor* scientists have been able to produce a detailed mathematical description of the Martian gravity

WHAT MAKES GRAVITY?

Gravity is among the least understood forces in nature. It is a fundamental attraction between all matter, but it is also a very weak force: The gravitational attraction of objects smaller than planets and moons is so weak that electrical or magnetic forces can easily oppose it. At the moment about the best that can be done with gravity is to describe its action: How much mass creates how much gravity? The question of what makes gravity itself is unanswered. This is part of the aim of a branch of mathematics and physics called string theory: to explain the relationships among the natural forces and to explain what they are in a fundamental way.

Sir Isaac Newton, the English physicist and mathematician who founded many of today's theories back in the mid-17th century, was the first to develop and record universal rules of gravitation. There is a legend that he was hit on the head by a falling apple while sitting under a tree thinking, and the fall of the apple under the force of Earth's gravity inspired him to think of matter attracting matter.

The most fundamental description of gravity is written in this way:

$$F = \frac{Gm_1m_2}{r^2},$$

where F is the force of gravity, G is the universal gravitational constant (equal to 6.67×10^{-11} Nm^2/kg^2), m_1 and m_2 are the masses of the two objects that are attracting each other with gravity, and r is the distance between the two objects. (N is the abbreviation for newtons, a metric unit of force.)

Immediately, it is apparent that the larger the masses, the larger the force of gravity. In addition, the closer together they are, the stronger the force of gravity, and because r is squared in the denominator, gravity diminishes very quickly as the distance between the objects increases. By substituting numbers for the mass of the Earth (5.9742×10^{24} kg), the mass of the Sun (1.989×10^{30} kg), and the distance between them, the force of gravity between the Earth and Sun is shown to be 8×10^{21} pounds per foot (3.56×10^{22} N). This is the force that keeps the Earth in orbit around the Sun. By comparison, the force of gravity between a piano player and her piano when she sits playing is about 6×10^{-7} pounds per foot (2.67×10^{-6} N). The force of a pencil pressing down in the palm of a hand under the influence of Earth's gravity is about 20,000 times stronger than the gravitational attraction between the player and the piano! So, although the player and the piano are attracted to each other by gravity, their masses are so small that the force is completely unimportant.

field. The Valles Marineris, a huge canyon and therefore a gravity low (the lack of mass from the canyon reduces gravity there, relative to a surface at the same height elsewhere) has a gravity anomaly reaching −450 mGals. The giant volcanic mountain Olympus Mons creates a gravity high, with a peak anomaly of 2,950 mGals. The Hellas impact basin has as a gravity low of −50 to −150 mGal, but its gravity structure is complicated, perhaps because of internal fracturing and shifting of the lithosphere at the time of the giant impact that created it.

Using the gravity and topography data together led researchers to conclude that the northern lowlands were likely a zone of high heat flow early in Martian history because of vigorous convection of the Martian interior. This rapid heat transfer could have released gases trapped within the planet to the atmosphere and underground ice or water to the surface, helping to produce a warmer, wetter climate than is present on Mars today.

REMNANTS OF AN ANCIENT MAGNETIC FIELD ON MARS

Imagine a freshly erupted lava flow, hot and liquid. As it cools, minerals begin to crystallize. One of these minerals may be magnetite (Fe_3O_4). While the mineral is still hot, it gains its own internal magnetic field in response to the planet's (or any other local environmental field). When it is completely cool, the grains of magnetite with the now solid lava retain their magnetization. This is called remanent magnetism, and in this way, rocks can retain through geologic time a record of the planet's magnetic field at the time they were formed. These rocks retain a record of the magnetic fields of the past, even if the present magnetic field has changed, or in the case of Mars, gone away entirely.

While magnetite is the most commonly mentioned mineral capable of holding a significant internal magnetic field from remanent magnetism, there are additional minerals that can do the same: pyrrhotite (Fe_7S_8), titanomagnetite (varying iron and titanium content from Fe_2O_3 to $FeTiO_3$), hematite (Fe_2O_3), and maghemite, which is a different crystal structure of the same composition as hematite. Other common rock-forming minerals, such as quartz and feldspar, do not retain a magnetic field. Most of the magnetic minerals listed here have been detected on Mars, either through orbital spectroscopy, surface analysis by landers, or in meteorites. All of these minerals also require strong oxidation to form, probably requiring water.

Now, imagine heating the rock again. At some temperature, while the magnetic mineral grains are still solid, they lose their magnetization. This is called the Curie temperature: the temperature above which a mineral loses its magnetism. Each kind of magnetic mineral has a different Curie temperature. For magnetite the temperature is about 1,100°F (600°C). In this way the record of the planet's past magnetic field can be erased, through volcanic heating, lightning strikes, or any other process that heats the minerals above their respective Curie temperatures. The Curie temperature is also reached inside the planet, where the planet is still naturally hot from the heat of formation and radiogenic heat. It is thought that the Curie temperature of 1,100°F (600°C) is reached at a depth of about 90 miles (150 km) in Mars today.

Currently Mars's magnetic field is about four-thousands as strong as that of Earth. This negligibly small field suggests that Mars has no convecting liquid core in the present day. (See the sidebar "What Makes Planetary Magnetic Fields?" on page 88.) The first evidence for a strong Martian magnetic field in the past was found in small crystals in Martian meteorites that carried a remanent magnetic field of their own. The magnetization of some of these tiny crystals suggested to some researchers that the Martian magnetic field was once at least within a factor of 10 of the strength of Earth's. Magnetic crystals of this type were found in the Martian meteorite Allen Hills 84001 (ALH 84001), and their strength suggested a magnetic field about 10 times smaller than Earth's, at 4.5 billion years ago when the meteorite's rock was formed.

Scientists have known for some time that Mars is no longer generating an internal magnetic field, but before *Mars Global Surveyor* there had been no opportunity to measure remanent magnetization in the Martian crust. *Mars Global Surveyor* measured magnetic fields on Mars globally and in detail and found some strongly magnetized crustal regions. The magnetization detected was recorded in the rocks when they were made, usually billions of years ago.

Several broad, intense alternating zones of magnetism were found in Terra Cimmeria in the southern hemisphere, and several small isolated anomalies were found in the north. The broad, alternating zones in the south resemble the alternating magnetic field stripes left on oceanic crust on Earth, as crust is formed at and moves away from the oceanic spreading centers, taking on the magnetic field of the Earth at that time, and recording the reversing magnetic fields of Earth over time. (The north magnetic pole of the Earth has switched directions repeatedly over Earth's history, pointing out of the South Pole of the Earth, and then switching back, over periods lasting millions of years.) The resemblance of the Martian stripes to the Earth's oceanic crustal stripes has lead some researchers to suggest that this section of the Martian crust formed in a similar way, spreading from an internal upwelling and recording a repeatedly reversing Martian magnetic field. Some researchers think that the planet had an active magnetic field only at the very beginning of the planet's history, in what is called the Noachian. Others think the dynamo and its magnetic field were more recent, after the Late Heavy Bombardment.

The theory of magma ocean crystallization and subsequent overturn of the mantle cumulates to a gravitationally stable configuration also provides a possible source for the Martian magnetic field. When dense, shallower cumulates flow down to the core-mantle boundary, they replace lighter cumulates that were also much hotter because of their initial depth. The dense cumulates form a colder layer around the core. The temperature difference between the core and these lowest cumulates creates a sudden, strong heat loss out of the core and into the cool cumulates. High heat flow is the cause of convection in the outer core, and heat flow caused by cumulate overturn may be sufficient to create a brief, intense Martian magnetic field early in the planet's history.

The southern hemisphere contains almost all the rocks with a significant magnetic signature. Some researchers think this indicates that the northern lowlands were created after the planetary magnetic field had died away, but in fact there is a way that the northern lowlands may have started out magnetized and then lost their magnetization. On Earth, ocean water circulates through cracks in the oceanic crust. The seawater

WHAT MAKES PLANETARY MAGNETIC FIELDS?

There is a prevalent idea for why terrestrial planets have magnetic fields, and though this idea is fairly well accepted by scientists, it is not thoroughly proved. On the Earth, it is known clearly that the outer core is liquid metal, almost entirely iron but with some nickel and also some small percentage of lighter elements (the exact composition is a topic of great argumentation at the moment in the scientific community). A planet's liquid outer core convects like boiling oatmeal around the solid inner core, and the moving currents of metal act like electrical currents, creating a magnetic field around them. The electrical to magnetic transition is simple, because every electrical current creates a magnetic field spiraling around itself; even the electrical wires in your house have magnetic fields around them. Electricity and magnetism are inseparable forces and always exist together.

The more innovative part of the theory concerns the convection of the liquid core: The liquid core is moving because it is carrying heat away from the inner core of the planet. Heat from the inner core conducts into the lowest part of the outer core, and that heated material expands slightly from the energy of the heating. Because it has expanded a tiny fraction it is now less dense than the unheated liquid next to it, and so it begins to rise through the radius of the outer core. When the heated material reaches the boundary with the mantle, it loses its heat by conducting it to the cooler mantle, and so it loses its extra buoyancy and sinks again. Many packages of material are going through this process of convection all at once, and together they form complex currents. It is common to describe convecting liquid as boiling oatmeal, but the liquid in the outer core is thought to be much less viscous (in other words, much more able to flow, much "thinner") than oatmeal. The convective currents are therefore thought to be much more complex and to be moving very rapidly.

Sir Joseph Larmour, an Irish physicist and mathematician, first proposed the hypothesis that the Earth's active magnetic field might be explained by the way the moving fluid iron in Earth's outer core mimics electrical currents, and the fact that every electric cur-

heats up by circulating through conduits in young, hot oceanic crust that has just formed from volcanism at a midocean ridge. Hot water is a very efficient solute, that is, it dissolves rock-forming minerals relatively quickly. The crystals break

rent has an associated, enveloping magnetic field. The combination of convective currents and a spinning inner and outer core is called the dynamo effect. If the fluid motion is fast, large, and conductive enough, then a magnetic field can not only be created but also carried and deformed by the moving fluid (this is also what happens in the Sun). This theory works well for the Earth and for the Sun, and even for Jupiter and Saturn, though they are thought to have a dynamo made of metallic hydrogen, not iron.

The theory is more difficult, potentially, for Mercury. Convection in the outer core cannot go on indefinitely because it is the process of moving heat out of the inner core and into the mantle, and then the mantle moves the heat out of the surface of the planet and into space. Over time, the inner, solid core grows as heat is lost from the outer core and more of it freezes into a solid, and eventually the planet has cooled enough that the dynamo is no longer active, and the planet no longer has a magnetic field. The Earth's field is still very active. Strong magnetism remaining in Mars's crust shows that it once had a magnetic field, but now the planet has none.

Mercury, much smaller than Mars, still has a magnetic field. Many scientists think it should have cooled completely by now, and therefore a dynamo in the outer core is impossible. In addition, when heat is moved from the outer core into the mantle, the mantle may be heated enough that it, too begins to convect. When the mantle convects, it can move hot material close enough to the surface that it melts through pressure release, and this creates volcanic eruptions on the planet's surface. As far as can be told from the photos from *Mariner 10* and from radar imaging from Earth, Mercury has not been volcanically active for a long time; in fact, its volcanic activity seems to have stopped even before the Moon's did. Together, Mercury's small size and apparent lack of volcanic activity argue against a magnetic field, and yet it has one. The exceptional size of its core may contribute to its continued field, but no one has any supporting data for different theories at the moment. Mercury's magnetic field is one of the major mysteries that the MESSENGER mission hopes to solve.

down into new minerals and lose parts of their composition into solution in the water. When iron-bearing minerals react with hot water, they lose their magnetization. It is possible, if the northern lowlands on Mars once held an ocean, that circulating ocean waters demagnetized the northern crust.

From the timing of the earliest formation of the core of Mars to the current state of the magnetic field, there are many strong hypotheses for the formation and evolution of Mars. All the models and hypotheses concerning Mars in the past and the interior of Mars have to rely on indirect information, such as the compositions of magmas that melted from the interior and erupted onto the surface, or the measurements of magnetic fields made by orbiting satellites. When these models and hypotheses match with direct observations scientists begin to be hopeful that the hypotheses may be going in the right direction. The latest Mars missions are providing floods of data from direct surface measurements, from temperatures, pressures, and surface compositions to rock types and the shapes of sedimentary strata. This exciting and new Martian data is described in the next chapter.

The Visible Planet

Images from Mars might make it look like a Utah desert on a hazy day. On the other hand, the surface of Mars is a less hospitable place. The average atmospheric pressure at the surface is about 1/100 of Earth's. Over the three years that the *Viking* lander measured atmospheric pressure, the pressure varied from just under 0.007 bars to just over 0.01 bars on an annual cycle that is driven by the northern hemisphere's surface temperature. The average surface temperature on Mars is −67°F (−55°C), though variations can be large. The *Mars Global Surveyor* Thermal Emission Spectrometer has measured polar temperatures as low as −200°F (−130°C) and equatorial temperatures as high as 30°F (−1°C). (See the sidebar "Remote Sensing" on page 102.) The Martian atmosphere is so thin that it adds just five degrees to the planet's surface temperature through insulating effects.

Beyond its low atmospheric pressure and freezing temperatures, Mars has seasonal dust storms of stunning ferocity. Dust storms have been known to cover the entire surface of the planet for months at a time, as can be seen in the photo on page 92. These images were taken in series during the month of June 2001. Several local dust storms are seen in the first image, especially along the retreating margin of the seasonal

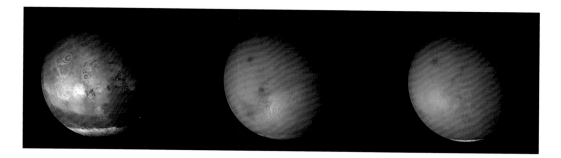

south polar frost cap and around the Hellas impact basin. Starting June 21 a small dust storm that began in Hellas grew larger, and overnight between June 25 and June 26 the storm's growth accelerated, and it soon spread over thousands of kilometers and then across the equator to become a global event.

The planet is also still volcanically active. Other volcanic flows, vents, and small volcanoes dot the planet. Several have frozen lava flows with fresh, uncratered surfaces, which planetary geologists think erupted within 20 million years of the present. Though 20 million years is unthinkably long to an individual person, it is practically considered current in the geological timescale. These observations, if they are correct, show that Mars is still volcanically active today. Mars seems, in fact, to have had regular volcanic eruptions over its entire 4.56-billion-year history, despite its complete lack of moving plates (most volcanic activity on Earth is related to plate tectonics). Most of the volcanic activity on Mars is centered on the huge volcanic complex known as Tharsis. Four gigantic volcanoes sit on the Tharsis rise, including Olympus Mons, 69,480 feet (21,183 m) high (almost three times the height Mauna Loa stands above the bottom of the Pacific Ocean).

SEASONAL ICE CAPS DRIVE ATMOSPHERE AND WEATHER

A planet's magnetic field protects the planet from the solar wind and therefore helps protect the atmosphere from being blown off and away (a strong gravity field is another good protector). Because Mars had only a brief, early magnetic field, it has been losing its atmosphere over time. It is estimated that

Mars has lost 95 to 99 percent of its atmosphere over the last 4 billion years. The Martian atmosphere now consists primarily of carbon dioxide (CO_2), with minor amounts of nitrogen (N_2) and other constituents. For a complete atmospheric composition, see the table below.

Using a thermal emission spectrometer (see sidebar "Remote Sensing" on page 102), the Mars Exploration Rover *Spirit* recently made a measurement of atmospheric composition by detecting the various wavelengths of infrared light emitted by molecules and particles in the atmosphere. This spectrum contains the signatures of carbon dioxide (wavelength of 15 microns), atmospheric dust (nine microns), and water vapor (six microns).

The first *Viking 1* color images showed a blue-gray sky, creating some excitement by its similarity to Earth. Fortunately a color test chart had been attached to the leg of the

MARTIAN ATMOSPHERIC COMPOSITION

Constituent	Chemical symbol	Fraction	Fraction on Earth
Carbon dioxide	CO_2	0.95	345 ppm
Nitrogen	N_2	0.027	0.7808
Argon	Ar	0.016	0.0093
Oxygen	O_2	0.013	0.2095
Carbon monoxide	CO	700 ppm	100 ppb
Water	H_2O	300 ppm	0.3 to 0.04 ppm
Neon	Ne	2.5 ppm	18 ppm
Krypton	Kr	0.3 ppm	1 ppm
Xenon	Xe	0.08 ppm	<1 ppt
Ozone	O_3	0.1 ppm	0 to 12 ppm

(Note: Ppm means parts per million; 1 ppm is equal to 0.000001 or 1×10^{-6}. ppb means "parts per billion": 1 ppb is equal to 0.000000001, or 1×10^{-9}. ppt means "parts per trillion": 1 ppt is equal to 0.000000000001, or 1×10^{-12}.

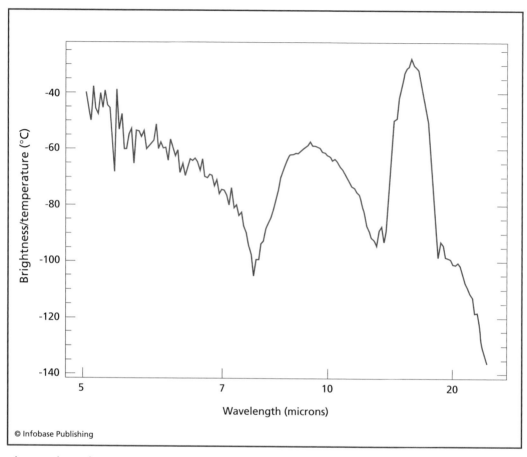

The mini-thermal emission spectrometer on the Mars Exploration Rover Spirit measured the wavelengths of infrared light emitted by the Martian sky, producing a spectrum that reveals the presence of specific molecules. The peak at 15 microns on this logarithmic scale is created by carbon dioxide, and the peak at 9 microns is diagnostic of atmospheric dust. (NASA/JPL/Arizona State University)

lander where it would appear in many photographs. Knowing the actual colors of the chart allowed scientists to correct the appearance of the photograph until the color chart was correct. These corrected images showed Mars's real sky color, an almost salmon-pink color caused by sunlight scattering off dust particles in the sky. Dust, in fact, is a major component of almost everything that happens on Mars: weather, clouds, atmosphere, erosion, and landforms.

Dust exists in the Martian atmosphere from the surface to about 30 miles (50 km) in height, but its density in the atmo-

These Martian clouds were photographed by the Mars Exploration Rover Opportunity on June 28, 2004. (NASA/ JPL/Malin Space Science Systems)

sphere is highly variable with time. The amount of dust in the atmosphere largely controls how clear the atmosphere is. Dust can create an optical depth as opaque as six (see the sidebar "Optical Depth" on page 96 for an explanation of its values) or as clear as 0.3. The atmosphere is thickest with dust, of course, after dust storms have lifted masses of dust into the atmosphere and left them hanging there.

Clouds on Mars, scanty though they are, consist primarily of dust, carbon dioxide, and water ice. Though there is relatively little water in the Martian atmosphere, it does exist at all heights between the surface and about 15 miles (25 km) height. In the winter at the poles there can even appear morning fog or thin and isolated water ice clouds. The water ice clouds, in fact, have a strong seasonal cycle. In northern summer, at Mars's aphelion, an equatorial cloud belt forms between 10 degrees south and 30 degrees north. Water ice clouds are common near high topography, however, during all seasons, particularly near the Tharsis rise, Alba Patera, and Elysium.

Carbon dioxide clouds are far more common, though they are thin (with an optical depth of about 0.001) and usually high, at about 38 miles (60 km). In the winter in polar regions, carbon dioxide clouds become much thicker (optical depth of about one) and lower, sometimes extending from the surface to about 15 miles (25 km) height.

Just as carbon dioxide and water dominate Martian clouds, so they also dominate the ice caps and frozen materials in the

OPTICAL DEPTH

Optical depth (usually denoted τ) gives a measure of how opaque a medium is to radiation passing through it. In the sense of planetary atmospheres, optical depth measures the degree to which atmospheric particles interact with light: Values of τ less than one mean very little sunlight is scattered by atmospheric particles or has its energy absorbed by them, and so light passes through the atmosphere to the planetary surface. Values of τ greater than one mean that much of the sunlight that strikes the planet's outer atmosphere is either absorbed or scattered by the atmosphere, and so does not reach the planet's surface. Values of τ greater than one for planets other than Earth also mean that it is hard for observers to see that planet's surface using an optical telescope.

Optical depth measurements use the variable z, meaning height above the planet's surface into its atmosphere. In the planetary sciences, τ is measured downward from the top of the atmosphere, and so τ increases as z decreases, so that at the planet's surface, τ is at its maximum, and z is zero. Each increment of τ is written as $d\tau$. This is differential notation, used in calculus, meaning an infinitesimal change in τ. The equation for optical depth also uses the variable κ (the Greek letter kappa) to stand for the opacity of the atmosphere, meaning the degree of light that can pass by the particular elemental makeup of the atmosphere. The Greek letter rho (ρ) stands for the density of the atmosphere, and dz, for infinitesimal change in z, height above the planet's surface.

$$d\tau = -\kappa \rho dz$$

Mathematical equations can be read just like English sentences. This one says, "Each tiny change in optical depth ($d\tau$) can be calculated by multiplying its tiny change in height (dz) by the density of the atmosphere and its opacity, and then changing

soils of the planet. Because of the eccentricity of Mars's orbit in the southern summer, Mars is much closer to the Sun. Southern summers are therefore hotter than northern summers on Mars, but also shorter, because Mars is moving fastest when it is closest to the Sun. In the cold season of each hemisphere, water ice and dry ice (carbon dioxide ice) form as seasonal polar caps on both the north and south poles. The white polar

the sign of the result" (this sign change is just another way to say that optical depth t increases as z decreases; they are opposite in sign).

To measure the optical depth of the entire atmosphere, this equation can be used on each tiny increment of height (z) and the results summed (or calculus can be used to integrate the equation, creating a new equation that does all the summation in one step). Optical depth also helps explain why the Sun looks red at sunrise and sunset but white in the middle of the day. At sunrise and sunset the light from the Sun is passing horizontally through the atmosphere, and thus has the greatest distance to travel through the atmosphere to reach an observer's eyes. At midday the light from the Sun passes more or less straight from the top to the bottom of the atmosphere, which is a much shorter path through the atmosphere (and let us remember here that no one should ever look straight at the Sun, since the intensity of the light may damage their eyes).

Sunlight in the optical range consists of red, orange, yellow, green, blue, indigo, and violet light, in order from longest wavelength to shortest (for more information and explanations, see appendix 2, "Light, Wavelength, and Radiation"). Light is scattered when it strikes something larger than itself, like a piece of dust, a huge molecule, or a drop of water, no matter how tiny, and bounces off in another direction. Violet light is the type most likely to be scattered in different directions as it passes through the atmosphere because of its short wavelength, thereby being shot away from the observer's line of sight and maybe even back into space. Red light is the least likely to be scattered, and therefore the most likely to pass through the longest distances of atmosphere on Earth and reach the observer's eye. This is why at sunset and sunrise the Sun appears red: Red light is the color most able to pass through the atmosphere and be seen. The more dust and water in the atmosphere, the more scattering occurs, so the more blue light is scattered away and the more red light comes through, the redder the Sun and sunset or sunrise appear.

caps can be seen clearly in photos throughout the Martian year, extending down many degrees in latitude during the hemisphere's winter and retreating back as temperatures rise.

On each pole there are also smaller, central ice caps that remain permanently throughout the year. The stable central portion of the northern polar cap is about three miles (five km) thick. These frozen caps contain a large volume of volatiles

Two Hubble Space Telescope images of Mars, taken about a month apart, reveal a state-sized dust storm churning near the edge of the Martian north polar cap. (NASA)

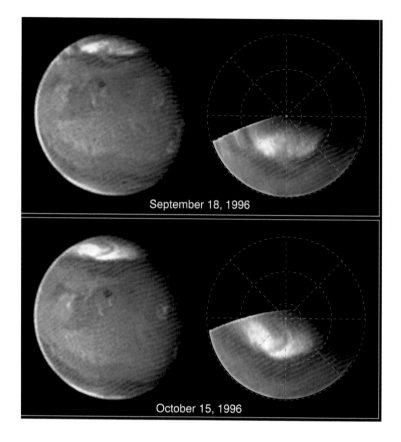

September 18, 1996

October 15, 1996

(chemicals that turn into liquid or gas at relatively low temperatures, in particular here, water and carbon dioxide). The volumes of volatiles in the ice caps on Mars are compared with those on Earth in the table on page 100. In high-resolution photos the ice caps on Mars can be seen to consist of many layers that show around the edges of the ice cap as shelves or benches, as shown in the accompanying figure. The streaks in the figure are dust devil tracks, and the image is about 1.9 miles (three km) in width.

Both polar caps consist primarily of carbon dioxide, but the north polar cap appears to contain more water ice than does the south pole cap. The caps contains two types of layers: Light-toned, nearly uniformly bedded layers lie above darker-toned beds that form shelves and benches at the bottom. The older, darker beds appear to include a large fraction of sand, while the upper layers may be a mixture of ice and dust.

Mars Reconnaissance Orbiter carries a high-resolution camera that has been revolutionizing understanding of the planet's surface. Many images have been taken of the north polar cap, which is now known to consist of many even layers of ice, one on top of the other like the pages of a book. The images show ~1 yard (1 m) per pixel of the image, so features must be larger than that size in order to be seen. Scientists have mapped and named the layers and are able to trace single layers for long

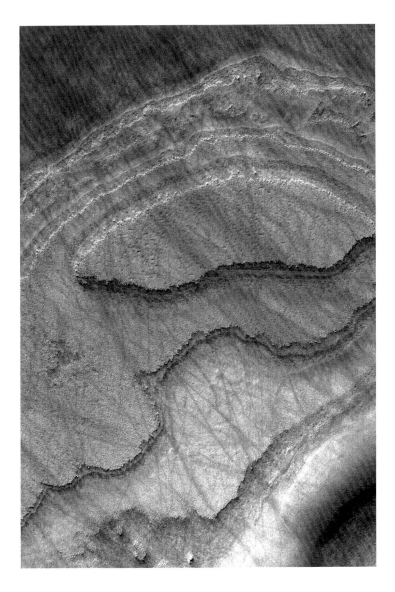

The side of a mesa in the south polar ice cap shows its layered internal structure. (NASA/JPL/ Malin Space Science Systems)

VOLUME OF VOLATILES ON THE MARTIAN SURFACE	
Reservoir	Volume (millions of cubic miles [km³])
Martian north polar cap	0.3 to 1.1 (1.2 to 1.7)
Martian south polar cap	0.5 to 0.7 (2 to 3)
Earth north polar cap	0.7 (3)
Earth south polar cap	7 (30)
Earth oceans	320 (1,300)

distances along the edges of the cap. The thick ice layers that can be recognized over distance are called marker beds and appear on average at intervals of about 18 yards (16 m) in the pile of ice.

In the Martian winter, when temperatures are approximately −195°F (−126°C), carbon dioxide and water freeze onto the polar cap. When winter turns to summer, a large portion of the ices sublime off the polar cap and return to the atmosphere, while some migrates to the other pole; over 62 lb/ft³ (1,000 kg/m³) of carbon dioxide ice moves out of a pole when its summer comes.

Recent missions to Mars, in particular the Thermal Emission Spectrometer on the *Mars Global Surveyor,* have collected a wealth of new data on Martian volatiles. (See the sidebar "Remote Sensing" on page 102.) This instrument was able to detect the sizes of carbon dioxide crystals and to measure the times of freezing and sublimation (transference from solid directly into the atmosphere). The Thermal Emission Spectrometer team detected fine-grained and coarse-grained carbon dioxide ice, as well as a larger, flatter shape they called slab ice. Condensation from gas to ice appears to happen on the surface of the ice caps rather than in the atmosphere, and the form of the ice can change over a matter of days. Data from this instrument could also be compared with data taken by the *Viking* mission 12 years previously, and little difference can be detected.

There is no place on Earth that carbon dioxide freezes into ice caps, and so the topography and behavior of carbon dioxide ice appears unusual to an Earth-based geologist. The image of the southern polar ice cap shows some of the odd topography that forms when carbon dioxide sublimes off the cap in southern summer. This image is about 1.2 miles square (2 km²). The mesa edges retreat about three meters each summer and refreeze from the atmosphere in winter.

(continues on page 108)

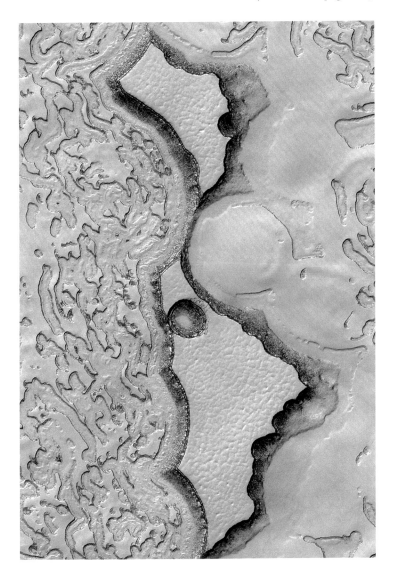

The carbon dioxide–rich south polar cap shows many round pits like the one in the center of this image. The pits grow quickly, showing that carbon dioxide is sublimating off the caps and into the atmosphere. Their flat bottoms indicate that they end at a layer, thought to be water ice, which is unable to sublime in the current temperatures. (NASA/ JPL/Malin Space Science Systems)

REMOTE SENSING

Remote sensing is the name given to a wide variety of techniques that allow observers to make measurements of a place they are physically far from. The most familiar type of remote sensing is the photograph taken by spacecraft or by giant telescopes on Earth. These photos can tell scientists a lot about a planet; by looking at surface topography and coloration photo geologists can locate faults, craters, lava flows, chasms, and other features that indicate the weather, volcanism, and tectonics of the body being studied. There are, however, critical questions about planets and moons that cannot be answered with visible-light photographs, such as the composition and temperature of the surface or atmosphere. Some planets, such as Venus, have clouds covering their faces, and so even photography of the surface is impossible.

For remote sensing of solar system objects, each wavelength of radiation can yield different information. Scientists frequently find it necessary to send detectors into space rather than making measurements from Earth, first because not all types of electromagnetic radiation can pass through the Earth's atmosphere (see figure, opposite page), and second, because some electromagnetic emissions must be measured close to their sources, because they are weak, or in order to make detailed maps of the surface being measured.

Spectrometers are instruments that spread light out into spectra, in which the energy being emitted at each wavelength is measured separately. The spectrum often ends up looking like a bar graph, in which the height of each bar shows how strongly that wavelength is present in the light. These bars are called spectral lines. Each type of atom can only absorb or emit light at certain wavelengths, so the location and spacing of the spectral lines indicate which atoms are present in the object absorbing and emitting the light. In this way, scientists can determine the composition of something simply from the light shining from it.

Below are examples of the uses of a number of types of electromagnetic radiation in remote sensing.

GAMMA RAYS

Gamma rays are a form of electromagnetic radiation; they have the shortest wavelength and highest energy. High-energy radiation such as X-rays and gamma rays are absorbed to a great degree by the Earth's atmosphere, so it is not possible to measure their production by solar system bodies without sending measuring devices into space.

Atmospheric Opacity

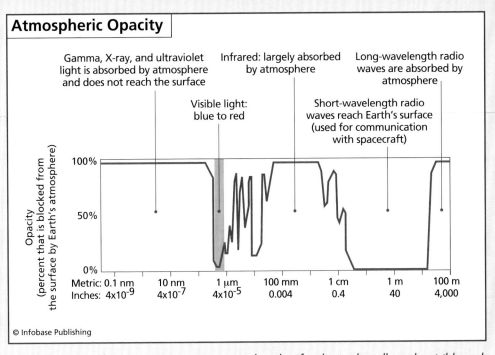

Gamma, X-ray, and ultraviolet light is absorbed by atmosphere and does not reach the surface

Infrared: largely absorbed by atmosphere

Long-wavelength radio waves are absorbed by atmosphere

Visible light: blue to red

Short-wavelength radio waves reach Earth's surface (used for communication with spacecraft)

Opacity (percent that is blocked from the surface by Earth's atmosphere)

100%

50%

0%

Metric: 0.1 nm   10 nm   1 μm   100 mm   1 cm   1 m   100 m
Inches: 4×10^{-9} 4×10^{-7} 4×10^{-5} 0.004   0.4   40   4,000

© Infobase Publishing

The Earth's atmosphere is opaque to many wavelengths of radiation but allows the visible and short radio wavelengths through to the surface.

These high-energy radiations are created only by high-energy events, such as matter heated to millions of degrees, high-speed collisions, or cosmic explosions. These wavelengths, then, are used to investigate the hottest regions of the Sun. The effects of gamma rays on other solar systems bodies, those without protective atmospheres, can be measured and used to infer compositions. This technique searches for radioactivity induced by the gamma rays.

Though in the solar system gamma rays are produced mainly by the hottest regions of the Sun, they can also be produced by colder bodies through a chain reaction of events, starting with high-energy cosmic rays. Space objects are continuously bombarded with cosmic rays, mostly high-energy protons. These high-energy protons strike the surface materials, such as dust and rocks, causing nuclear reactions in the atoms of the surface material. The reactions produce neutrons, which collide with surrounding nuclei. The

(continued)

nuclei become excited by the added energy of neutron impacts, and reemit gamma rays as they return to their original, lower-energy state. The energy of the resultant gamma rays is characteristic of specific nuclear interactions in the surface, so measuring their intensity and wavelength allow a measurement of the abundance of several elements. One of these is hydrogen, which has a prominent gamma-ray emission at 2.223 million electron volts (a measure of the energy of the gamma ray). This can be measured from orbit, as it has been in the Mars Odyssey mission using a Gamma-Ray Spectrometer. The neutrons produced by the cosmic ray interactions discussed earlier start out with high energies, so they are called fast neutrons. As they interact with the nuclei of other atoms, the neutrons begin to slow down, reaching an intermediate range called epithermal neutrons. The slowing-down process is not too efficient because the neutrons bounce off large nuclei without losing much energy (hence speed). However, when neutrons interact with hydrogen nuclei, which are about the same mass as neutrons, they lose considerable energy, becoming thermal, or slow, neutrons. (The thermal neutrons can be captured by other atomic nuclei, which then can emit additional gamma rays.) The more hydrogen there is in the surface, the more thermal neutrons relative to epithermal neutrons. Many neutrons escape from the surface, flying up into space where they can be detected by the neutron detector on Mars Odyssey. The same technique was used to identify hydrogen enrichments, interpreted as water ice, in the polar regions of the Moon.

X-RAYS

When an X-ray strikes an atom, its energy can be transferred to the electrons or biting the atom. This addition of energy to the electrons makes one or more electrons leap from their normal orbital shells around the nucleus of the atom to higher orbital shells, leaving vacant shells at lower energy values. Having vacant, lower-energy orbital shells is an unstable state for an atom, and so in a short period of time the electrons fall back into their original orbital shells, and in the process emit another X-ray. This X-ray has energy equivalent to the difference in energies between the higher and lower orbital shells that the electron moved between. Because each element has a unique set of energy levels between electron orbitals, each element produces X-rays with energies that are characteristic of itself and no other element. This method can be used remotely from a satellite, and it can also be used directly on tiny samples of material placed in a laboratory instrument called an electron microprobe, which measures the composition of the material based on the X-rays the atoms emit when struck with electrons.

VISIBLE AND NEAR-INFRARED

The most commonly seen type of remote sensing is, of course, visible light photography. Even visible light, when measured and analyzed according to wavelength and intensity, can be used to learn more about the body reflecting it.

Visible and near-infrared reflectance spectroscopy can help identify minerals that are crystals made of many elements, while other types of spectrometry identify individual types of atoms. When light shines on a mineral, some wavelengths are absorbed by the mineral, while other wavelengths are reflected back or transmitted through the mineral. This is why things have color to the eye: Eyes see and brains decode the wavelengths, or colors, that are not absorbed. The wavelengths of light that are absorbed are effectively a fingerprint of each mineral, so an analysis of absorbed versus reflected light can be used to identify minerals. This is not commonly used in laboratories to identify minerals, but it is used in remote sensing observations of planets.

The primary association of infrared radiation is heat, also called thermal radiation. Any material made of atoms and molecules at a temperature above absolute zero produces infrared radiation, which is produced by the motion of its atoms and molecules. At absolute zero, $-459.67°F$ ($-273.15°C$), all atomic and molecular motion ceases. The higher the temperature, the more they move, and the more infrared radiation they produce. Therefore, even extremely cold objects, like the surface of Pluto, emit infrared radiation. Hot objects, like metal heated by a welder's torch, emit radiation in the visible spectrum as well as in the infrared.

In 1879 Josef Stefan, an Austrian scientist, deduced the relation between temperature and infrared emissions from empirical measurements. In 1884 his student, Ludwig Boltzmann derived the same law from thermodynamic theory. The relation gives the total energy emitted by an object (E) in terms of its absolute temperature in Kelvin (T), and a constant called the Stefan-Boltzmann constant (equal to 5.670400×10^{-8} W m^{-2} K^{-4}, and denoted with the Greek letter sigma, σ):

$$E = \sigma T^4$$

This total energy E is spread out at various wavelengths of radiation, but the energy peaks at a wavelength characteristic of the temperature of the body emitting the energy. The relation between wavelength and total energy, Planck's Law, allows scientists to determine the temperature of a body by measuring the energy it emits.

(continues)

(continued)

The hotter the body, the more energy it emits at shorter wavelengths. The surface temperature of the Sun is 9,900°F (5,500°C), and its Planck curve peaks in the visible wavelength range. For bodies cooler than the Sun, the peak of the Planck curve shifts to longer wavelengths, until a temperature is reached such that very little radiant energy is emitted in the visible range.

Humans radiate most strongly at an infrared wavelength of 10 microns (*micron* is another word for micrometer, one millionth of a meter). This infrared radiation is what makes night vision goggles possible: Humans are usually at a different temperature than their surroundings, and so their shapes can be seen in the infrared.

Only a few narrow bands of infrared light make it through the Earth's atmosphere without being absorbed, and can be measured by devices on Earth. To measure infrared emissions, the detectors themselves must be cooled to very low temperatures, or their own infrared emissions will swamp those they are trying to measure from elsewhere.

In thermal emission spectroscopy, a technique for remote sensing, the detector takes photos using infrared wavelengths and records how much of the light at each wavelength the material reflects from its surface. This technique can identify minerals and also estimate some physical properties, such as grain size. Minerals at temperatures above absolute zero emit radiation in the infrared, with characteristic peaks and valleys on plots of emission intensity versus wavelength. Though overall emission intensity is determined by temperature, the relationships between wavelength and emission intensity are determined by composition. The imager for *Mars Pathfinder,* a camera of this type, went to Mars in July 1997 to take measurements of light reflecting off the surfaces of Martian rocks (called reflectance spectra), and this data was used to infer what minerals the rocks contain.

When imaging in the optical or near-infrared wavelengths, the image gains information about only the upper microns of the surface. The thermal infrared gives information about the upper few centimeters, but to get information about deeper materials, even longer wavelengths must be used.

RADIO WAVES

Radio waves from outside the Earth do reach through the atmosphere and can be detected both day and night, cloudy or clear, from Earth-based observatories using huge metal dishes. In this way, astronomers observe the universe as it appears in radio waves. Images like photographs can be made from any wavelength of radiation coming from a body: Bright regions on the image can correspond to more intense radiation,

Planck Curves for Black Bodies

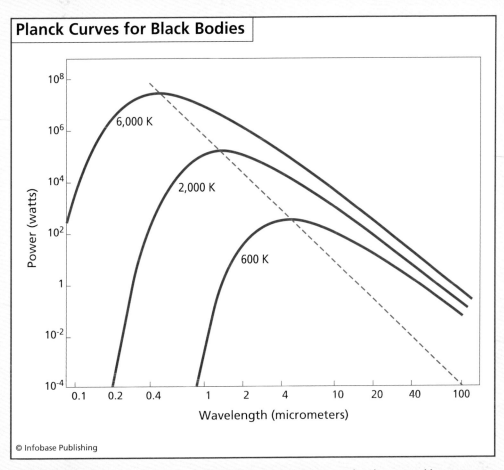

© Infobase Publishing

The infrared radiation emitted by a body allows its temperature to be determined by remote sensing; the curves showing the relationship between infrared and temperature are known as Planck curves.

and dark parts, to less intense regions. It is as if observers are looking at the object through eyes that "see" in the radio, or ultraviolet, or any other wavelength, rather than just visible. Because of a lingering feeling that humankind still observes the universe exclusively through our own eyes and ears, scientists still often refer to "seeing" a body in visible wavelengths and to "listening" to it in radio wavelengths.

(continues)

(continued)

Radio waves can also be used to examine planets' surfaces, using the technique called radar (radio detection and ranging). Radar measures the strength and round-trip time of microwave or radio waves that are emitted by a radar antenna and bounced off a distant surface or object, thereby gaining information about the material of the target. The radar antenna alternately transmits and receives pulses at particular wavelengths (in the range 1 cm to 1 m) and polarizations (waves polarized in a single vertical or horizontal plane). For an imaging radar system, about 1,500 high-power pulses per second are transmitted toward the target or imaging area. At the Earth's surface, the energy in the radar pulse is scattered in all directions, with some reflected back toward the antenna. This backscatter returns to the radar as a weaker radar echo and is received by the antenna in a specific polarization (horizontal or vertical, not necessarily the same as the transmitted pulse). Given that the radar pulse travels at the speed of light, the measured time for the round trip of a particular pulse can be used to calculate the distance to the target.

Radar can be used to examine the composition, size, shape, and surface roughness of the target. The antenna measures the ratio of horizontally polarized radio waves sent to the surface to the horizontally polarized waves reflected back, and the same for vertically polarized waves. The difference between these ratios helps to measure the roughness of the surface. The composition of the target helps determine the amount of energy that is returned to the antenna: Ice is "low loss" to radar, in other words, the radio waves pass straight through it the way light passes through window glass. Water, on the other hand, is reflective. Therefore, by measuring the intensity of the returned signal and its polarization, information about the composition and roughness of the surface can be obtained. Radar can even penetrate surfaces and give information about material deeper in the target: By using wavelengths of three, 12.6, and 70 centimeters, scientists can examine the Moon's surface to a depth of 32 feet (10 m), at a resolution of 330 to 985 feet (100 to 300 m), from the Earth-based U.S. National Astronomy and Ionosphere Center's Arecibo Observatory!

(continued from page 101)

Every southern spring, a dark region forms on the south polar cap. This dark region has been noticed since 1845 from Earth-based telescopes and since 1997 from Mars orbiters. Originally the southern polar cap was thought to consist of water ice and so the dark patch was assumed to be a lake that formed from melting in the spring. The new information from

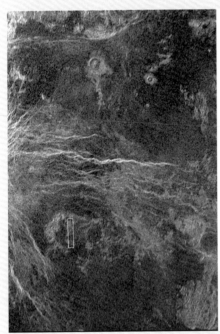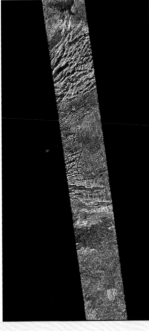

The far greater resolution obtained by the Magellan craft (right) shows the relative disadvantage of taking images of Venus from the Earth (left) using the Arecibo observatory. (NASA/ Magellan/JPL)

Venus is imaged almost exclusively in radar because of its dense, complete, permanent cloud cover. Radar images of Venus have been taken by several spacecraft and can also be taken from Arecibo Observatory on Earth. The image above makes a comparison between the resolution possible from Earth using Arecibo (left), and the resolution from the *Magellan* spacecraft (right). Arecibo's image is 560 miles (900 km) across and has a resolution of 1.9 miles (3 km). The *Magellan* image corresponds to the small white rectangle in the Arecibo image, 12 × 94 miles (20 × 120 km) in area. Magellan's resolution is a mere 400 feet (120 m) per pixel.

missions shows that the southern polar cap is mostly carbon dioxide ice with only a very small fraction of water and that the dark spot is a very complex collection of patterns and features in the ice, subliming into gas in the southern spring. The dark area contains fields of dark fan-shaped spots, areas with ragged branching dark patterns shaped like spiders, and fields of spots surrounded with halos. Three of these regions

The Martian north polar cap contains frozen water but is difficult to photograph because it spends six months in darkness every winter and is often obscured by dust storms in the summer. (NASA/JPL/ Malin Space Science Systems)

have been named informally by researchers working with *Mars Reconnaissance Orbiter* data: Manhattan, Ithaca, and Giza. The current theory of the researchers is that the dark areas are caused by ice loss from spring winds across the ice cap.

The action of heating and cooling with the seasons and freezing and sublimating carbon dioxide at the Martian poles controls much of the air movement on Mars. In the northern hemisphere's spring a high-pressure system is created by the subliming carbon dioxide thickening the atmosphere. At the same time the south pole is moving into winter and carbon dioxide is freezing out of the atmosphere, making the atmosphere thinner and creating low-pressure systems. This difference in pressure between the two hemispheres results in air moving across the planet from north to south. Half a year later, in northern autumn and southern spring, air moves in the same manner but in the opposite direction: from south to north.

The volumes of material moving into and out of the poles each season is immense: Up to 25 percent of Mars's entire atmosphere freezes out each season into the polar caps and the atmospheric pressure of the planet declines by a similar percentage. The result of these large-scale carbon dioxide exchanges is that weather systems on Mars tend to be global rather than local, as they are on Earth.

In winter, the southern ice cap extends to 50 degrees south and is one to two meters thick around its central permanent area; in summer the cap remains as a small residual core. Because of the shortness of the southern summer, the southern cap retains some frozen carbon dioxide as well as frozen water.

A characteristic feature of the north polar cap is a spiraling pattern of scarps and troughs organized around the pole. Horizontal or north-facing areas appear white, while the scarps expose dark layers. The spiraling scarps and troughs are thought to be formed by a combination of sublimation, wind effects, deposition, and ice flow. The Mars Orbital Laser Altimeter topographic data has been used for an ice sheet model by Christine Schott Hvidberg, of the University of Copenhagen, Denmark, to study these mechanisms, their effects and relative importance under the present climatic conditions. The results suggest that the spiraling structure is formed by sublimation combined with wind effects. Ice flow alone would close the troughs within 100,000 to 1 million years. Hvidberg has estimated the total amount of sublimation to be in the order of 10^{11} to 10^{12} kg per Martian year.

Through the extremity of its seasons and its limited volatile budget, then, Mars's atmospheric circulation, composition, and pressure are all controlled by the freezing and sublimation of carbon dioxide and, to a lesser extent, water, at its poles.

The great differences in atmospheric pressure and winds between the seasons caused by sublimation and freezing of the ice caps create large weather patterns. In 1977, the *Viking* lander survived a giant dust storm that put so much dust into the atmosphere that sunlight reaching the planet's surface was reduced to between 1 and 5 percent. This and other giant dust storms are caused by the intense heating of southern summer and the large amount of ice sublimed as a result in the short period of perihelion, when Mars is moving the fastest past the Sun. These giant dust storms commonly grow to cover the entire planet and carry winds as high as 250 miles per hour (400 km/hr). The high winds of these pressure systems create the spiraling erosion patterns at the poles. The dust carried by high winds is a strong erosive force on the rest of the planet as well, carving surface rocks into distinctive streamlined shapes called yardangs.

Mars's atmosphere warms when dust storms are active. Recent orbital missions have observed the life cycles of five large dust storms and have measured warming of 15 degrees in the atmosphere where the storms are active. The warmer atmosphere removes the water clouds from the atmosphere.

MAPPING THE SURFACE OF MARS

Mapping the surface of Mars has a long and convoluted history consisting not only of changing observations made through increasingly powerful telescopes but misinterpretations of surface features in eager attempts to prove the existence of life on another planet. The belief in life on Mars may have begun with a mistranslation from Italian to English. In the late 19th century Angelo Secchi and Giovanni Schiaparelli, a pair of Italian astronomers, published detailed maps of Mars based on their observations through telescopes. Secchi, a Jesuit monk, drew his map in 1858, calling Syrtis Major the "Atlantic Canale." In 1863 he made further color sketches of Mars and referred to channels as "canali." In both of these cases the Italian word *canale,* and its plural, *canali,* mean channels or grooves, in other words, linear features. When Schiaparelli drew his own map of Mars in 1877, using names from historic and mythological sources, he used Secchi's terms *canale* and *canali* to refer to a network of dark and light lines. "Canali" was mistranslated as "canals," taken by some to imply engineering by intelligent life on the Martian surface (the recent completion of the Suez Canal, in 1869, may have made people quicker to assume that canals were built by intelligent life). This had not been the implication of Schiaparelli at all, who had simply used the word *canali* to indicate linear features, but the mistranslation was the major trigger of a mania for the search for Martians that persists to this day.

Earlier scientists had also proposed that there was life on Mars. In 1854 William Whewell, the great British historian and philosopher of science, speculated about life on Mars; his speculations received attention because of his prominence in science. Whewell invented the terms *anode, cathode,* and *ion* for Michael Faraday, who pioneered the understanding of electricity and magnetism and was perhaps the finest experimental scientist who ever lived. Upon the request in 1833 of Samuel Taylor Coleridge, the great English poet, Whewell invented the English word *scientist.* Before this time the only terms in use were *natural philosopher* and *man of science.* In 1860 Emmanuel Liais, deputy manager of the Observatoire de

Paris, proposed that the dark regions on Mars were not seas but vegetation. The mistranslation of "canali" in 1877 was the event that brought the powerful American scientist Percival Lowell into the search for Martian life. Lowell's brother was the president of Harvard University, and later in his life, Lowell himself was a professor at the Massachusetts Institute of Technology.

In 1894 Lowell built a huge observatory in Flagstaff, Arizona, with a 24-inch telescope, and named it the Lowell Observatory. He hired scientists to look for the canals and, therefore, evidence for life on Mars. Lowell also spent much of his time studying linear features on Mars and the seasonal changes in polar caps, and he estimated that Martian surface temperature averaged 48°F (9°C). He drew complex maps of the canals he perceived, one of which he claimed was 3,500 miles (5,630 km) long (the longest canal on Earth is 745 miles, or 1,200 km long). He stated that the canals radiated from the polar caps, long assumed to be made of water ice, and converged at dark spots, which he inferred to indicate that the Martians pumped water from the poles to cities. Lowell agreed with Liais and others who thought certain green patches were plants, and he also stated that the Martians had learned to live in global peace, as evidenced by their global canal system.

Lowell published his views in three books: *Mars* (1895), *Mars and Its Canals* (1906), and *Mars As the Abode of Life* (1908). Plans were devised for signaling the Martians, despite the obvious problems with translation. There were proposals to make a network of mirrors to flash messages to Mars, and another proposal to fill a 20-mile long trench in the Sahara with kerosene and set it alight. H. G. Wells's book *The War of the Worlds,* a wonderful science fiction work about an attack on Earth by Martians, was serialized in *Pearson's Magazine* during 1897 and printed in hardback the following year. From that point onward, the notion of little green men from Mars was cemented into popular culture.

There was really one large problem with all the theories about life on Mars, which centered on the observation of canals: Lowell was the only person who could make consistent observations of them. Other astronomers could neither see nor photograph the elusive canals. A number of prominent scientists of the time were skeptical of the Mars theory, including E. E. Barnard at Lick Observatory, E. W. Maunder in England, Charles A. Young at Princeton, Asaph Hall at the U.S. Naval Observatory, and G. E. Hale, founder of Mount Wilson Observatory, because the observations could not be reproduced. Hale, using the largest reflecting telescope in the world at one of the finest sites for image stability, described in 1909 that the Martian surface markings mistaken for straight lines "were vague and diffuse and by no means narrow and sharp" and were "made up of interlacing and curved filaments." In fact in the end this proved to be the case: The only features to be seen are curving ancient river valleys, the edges of lava flows, and sand dune fields, all natural features. It seems the canals were really dark patches and other surface features, that when seen through a low-resolution telescope, seem to blur together into lines, though this may still be giving the benefit of the doubt.

Even the *Encyclopædia Britannica,* 11th Edition (1910) doubts their being waterways, noting that their breadth of many miles made it absurd to call them canals but believed that the oft-noted seasonal changes were evidence of blue-green vegetation. Lowell maintained that the linear features were indeed of artificial origin, and the *Britannica* quotes him:

Professor Lowell's theory is supported by so much evidence of different kinds that his own exposition should be read *in extenso* in *Mars and its canals* (1906) and *Mars as the abode of life* (1909). In order, however, that his views may be adequately presented here, he has kindly supplied the following summary in his own words: "Owing to inadequate atmospheric advantages generally, much misapprehension exists as to the definiteness with which the surface of Mars is seen under good conditions. In steady air the canals are perfectly distinct lines, not unlike the Fraunhofer ones of the Spectrum, pencil lines or gossamer filaments according to size. All the observers at Flagstaff concur in this. The photographs of them taken there also confirm it up to the limit of their ability . . . differences in drawings are differences of time and are due to seasonal and secular changes in the planet itself. These seasonal changes have been carefully followed at Flagstaff, and the law governing them detected. They are found to depend upon the melting of the polar caps. After the melting is under way the canals next to the cap proceed to darken, and the darkening thence progresses regularly down the latitudes. Twice this happens every Martian year, first from one cap and then six Martian months later from the other. . . ." These facts and a host of others of like significance have led Lowell to the conclusion that the whole canal system is of artificial origin, first because of each appearance and secondly because of the laws governing its development. The warmer temperature disclosed from Lowell's investigation on the subject, and the spectrographic detection by Slipher of water-vapour in the Martian air, are among the latest of these confirmations.

Unfortunately, Lowell's descriptions were flawed. Vesto Melvin Slipher's methods were not sensitive enough to detect atmospheric water vapor. Contemporary astronomers Eugène-Michel Antoniadi and Hale disputed the geometrical patterns that he termed canals, and few believed them to be waterways, though the myth has been maintained in popular writing. Sadly, Lowell also dedicated a huge amount of time to finding Planet X, the suspected planet beyond Neptune, but he never found it, despite the fact that it has since been shown that he actually had photographed it but failed to recognize it!

The surface of Mars is mapped by photographs at varying scales of resolution and thoroughly mapped by gravity, temperature, and topographic measurements. The laser altimeter measurements have mapped altitudes all over Mars so completely and accurately that the surface of Mars is better known than the surface of the Earth and is much better known than the surface of the Moon (the altitudes of features on the Moon near the poles are only known to within about 1.2 miles [~2 km]). All this data can be taken together to learn about Mars today and also to project into the Martian past. The table on page 116 lists some elevations on Mars, including the largest volcano in the solar system (Olympus Mons) and one of the largest valley systems in the solar system (Valles Marineris).

VOLCANOES

Mars has an obvious volcanic past, seen both in the large volcanoes prominent on its surface and in the volcanic rock that makes up the meteorites that have reached Earth. The planet's volcanic history spans the entire age of the planet itself. Studies of age relations show that the giant Tharsis volcanic rise formed before 4 billion years ago, in fact within the first few hundred million years of the planet's existence. Volcanic Martian meteorites formed between 1.3 billion years ago and just a few million years ago, according to dating using radionuclides. (See the sidebar "Determining Age from Radioactive Isotopes" on page 39.) By studying relationships among flows on the surface of the planet, some geologists believe that the most recent volcanic eruption was as recent as 10 million years ago. The planet is almost certainly still volcanically active.

SELECTED ELEVATIONS ON MARS	
Feature	Elevation feet (m)
Olympus Mons	69,808 (21,283)
Ascraeus Mons	59,699 (18,201)
Arsia Mons	57,085 (17,404)
Pavonis Mons	46,320 (14,122)
Elysium	43,587 (13,289)
Alba Patera	21,713 (6,620)
lowest point in Valles Marineris	−17,416 (−5,310)
Hellas Basin	−25,666 (−7,825)

(Note: Olympus Mons has been widely reported as 17 miles [27 km] in height, but the final MOLA data gives the lower value given here.)

The Tharsis rise is one of the most prominent features on the surface of Mars. Tharsis is a huge topographic rise that holds four volcanoes, including the largest Martian shield volcanoes. The Tharsis ridge stands 5.5 miles (9 km) high and lies on the Martian equator. These volcanoes are about 250 miles (400 km) in diameter and reach elevations of 10.6 miles (17 km).

The summit calderas (central depressions) of all four volcanoes probably formed from recurrent collapse following drainage of magma resulting from flank eruptions. Alba Patera in the north center of the image, 1,000 miles (1,600 km) in diameter, far exceeds any other volcano in the solar system in area, covering eight times the area of Olympus Mons but reaching only about 3.8 miles (6 km) in height.

Olympus Mons (below), the largest volcano on Mars and in fact in the entire solar system, lies just to the west of the Tharsis rise. Olympus Mons is 375 miles (600 km) in diameter and 13.16 miles (21,283 m) in height (this volcano has been widely reported as 17 miles or 27 km in height, but the final MOLA data gives the lower value given here). By comparison, the Hawaiian island of Mauna Loa stands only five miles (8 km) above the floor of the Pacific Ocean. The caldera on Olympus Mons (the circular depression in the volcano's peak from which recent eruptions have come) is 50 miles (80 km)

This image of Olympus Mons, the largest volcano in the solar system, was created by adding topographic data to photographic images— the vertical exaggeration is 10:1. (NASA/MOLA Science Team)

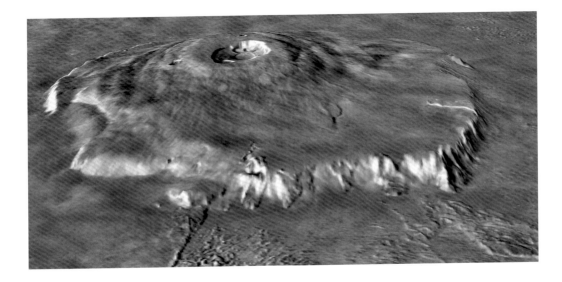

across. Olympus Mons was called Nix Olympica ("the Snows of Olympus") by its first telescopic observers, who could only see a white feature. When telescopes improved to the point that they could distinguish the feature as a mountain, it was renamed Olympus Mons.

The image of Olympus Mons on page 117 is processed to combine the photographic image of the volcano with topographic (elevation) information from the Mars Orbital Laser Altimeter, producing an image that shows appearance and height. Olympus Mons is surrounded by a well-defined cliff that is up to 3.5 miles (6 km) high. At its highest point, this cliff is more than three times as high as the Grand Canyon is deep.

The great weight and height of the Tharsis region causes stress in the lithosphere, the outer, rigid shell of the planet, much as a person sitting on a bed presses down the mattress and causes it to bend. Faults and ridges surround Tharsis.

Mars's lithosphere must be very thick and stiff to support the great height and weight of Tharsis through most of Mars's history: It has been shown that Tharsis must have formed in the Noachian epoch, very early in Martian history. This is evidence, therefore, that Mars's lithosphere was thick and stiff even then, or Tharsis would have collapsed before now. Whatever theory is called on to explain Tharsis, then, has to work on a planet that already has a very thick, stiff lithosphere, and that is probably already a one-plate planet (any plate tectonic activity that had occurred would have had to be over, and the planet was covered with a single-plate shell).

Wrinkle ridges are globally distributed on Mars, but concentric to Tharsis, the huge volcanic center. Researchers believe that the plains formed first, and then the wrinkle ridges. It has been suggested that wrinkle ridges formed from cooling, either long-term cooling or cooling after volcanism. When solid materials cool, they generally contract. When the interior of a planet cools and contracts, the surface area is then slightly too large to cover the newly contracted interior, and it may then form wrinkles. Wrinkle ridges may also have been formed by warping the surface of the planet by loading it with ice, as in large glaciers, or with volcanic products, like

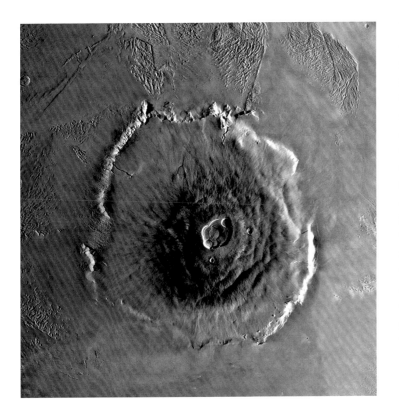

The central edifice of Olympus Mons has a summit caldera 38 miles (24 km) above the surrounding plains. Surrounding the volcano is an outward-facing scarp 880 miles (550 km) in diameter and about a mile (several kilometers) high. Beyond the scarp is a moat filled with lava and material that fell as landslides off the sharp scarp. Farther out is an aureole of characteristically grooved terrain, just visible at the top of the frame. (NASA/NSSDC)

Tharsis. It is currently thought that only the loading of Tharsis provides the amount of strain observed in the number and size of the wrinkle ridges.

The Tharsis region has also created an odd phenomenon called perched or pedestal craters. These craters are normal impact craters, but they are raised from the surrounding surface. The crater bottom is at a higher elevation than the plain the crater lies upon. There is no cratering mechanism that can cause this; cratering commonly creates a crater bottom that is at a lower elevation than the surroundings when the crater is excavated. The elevations of the crater bottoms have been measured and found to decrease smoothly with distance from Tharsis. The current model for perched craters calls on an immense ash fall from Tharsis that blankets all the surrounding plains with their craters. Later, the violent storms of Mars sweep over the plains and scour away all the ash and regolith

(continues on page 122)

FOSSA, SULCI, AND OTHER TERMS FOR PLANETARY LANDFORMS

On Earth the names for geological features often connote how they were formed and what they mean in terms of surface and planetary evolution. A caldera, for example, is a round depression formed by volcanic activity and generally encompassing volcanic vents. Though a round depression on another planet may remind a planetary geologist of a terrestrial caldera, it would be misleading to call that feature a caldera until its volcanic nature was proven. Images of other planets are not always clear and seldom include topography, so at times the details of the shape in question cannot be determined, making their definition even harder.

To avoid assigning causes to the shapes of landforms on other planets, scientists have resorted to creating a new series of names largely based on Latin, many of which are listed in the following table, that are used to describe planetary features. Some are used mainly on a single planet with unusual features, and others can be found throughout the solar system. Chaos terrain, for example, can be found on Mars, Mercury, and Jupiter's moon Europa. The Moon has a number of names for its exclusive use, including lacus, palus, rille, oceanus, and mare. New names for planetary objects must be submitted to and approved by the International Astronomical Union's (IAU) Working Group for Planetary System Nomenclature.

NOMENCLATURE FOR PLANETARY FEATURES

Feature	Description
astrum, astra	radial-patterned features on Venus
catena, catenae	chains of craters
chaos	distinctive area of broken terrain
chasma, chasmata	a deep, elongated, steep-sided valley or gorge
colles	small hills or knobs
corona, coronae	oval-shaped feature
crater, craters	a circular depression not necessarily created by impact
dorsum, dorsa	ridge
facula, faculae	bright spot

Feature	Description
fluctus	flow terrain
fossa, fossae	narrow, shallow, linear depression
labes	landslide
labyrinthus, labyrinthi	complex of intersecting valleys
lacus	small plain on the Moon; name means "lake"
lenticula, lenticulae	small dark spots on Europa (Latin for freckles); may be domes or pits
linea, lineae	a dark or bright elongate marking, may be curved or straight
macula, maculae	dark spot, may be irregular
mare, maria	large circular plain on the Moon; name means "sea"
mensa, mensae	a flat-topped hill with cliff-like edges
mons, montes	mountain
oceanus	a very large dark plain on the Moon; name means "ocean"
palus, paludes	small plain on the Moon; name means "swamp"
patera, paterae	an irregular crater
planitia, planitiae	low plain
planum, plana	plateau or high plain
reticulum, reticula	reticular (netlike) pattern on Venus
rille	narrow valley
rima, rimae	fissure on the Moon
rupes	scarp
sinus	small rounded plain; name means "bay"

(continues)

(continued)

NOMENCLATURE FOR PLANETARY FEATURES (continued)

Feature	Description
sulcus, sulci	subparallel furrows and ridges
terra, terrae	extensive land mass
tessera, tesserae	tile-like, polygonal terrain
tholus, tholi	small dome-shaped mountain or hill
undae	dunes
vallis, valles	valley
vastitas, vastitates	extensive plain

The IAU has designated categories of names from which to choose for each planetary body, and in some cases, for each type of feature on a given planetary body. On Mercury, craters are named for famous deceased artists of various stripes, while rupes are named for scientific expeditions. On Venus, craters larger than 12.4 miles (20 km) are named for famous women, and those smaller than 12.4 miles (20 km) are given common female first names. Colles are named for sea goddesses, dorsa are named for sky goddesses, fossae are named for goddesses of war, and fluctus are named for miscellaneous goddesses.

The gas giant planets do not have features permanent enough to merit a nomenclature of features, but some of their solid moons do. Io's features are named after characters from Dante's *Inferno*. Europa's features are named after characters from Celtic myth. Guidelines can become even more explicit: Features on the moon Mimas are named after people and places from Malory's *Le Morte d'Arthur* legends, Baines translation. A number of asteroids also have naming guidelines. Features on 253 Mathilde, for example, are named after the coalfields and basins of Earth.

(continued from page 119)

from around the craters, but the crater bottoms with their loads of ash are relatively protected from the winds by the crater walls. The ash layers in the craters reflect the thickness of the ash fall from Tharsis and its natural decline with distance, while the plains around the craters have been scoured clean, leaving the craters perched above the new, lower plain level.

There is, at this time, no really convincing theory for the formation of Tharsis. There is no analog on any other terrestrial planet. In some theories, a relatively narrow hot upwelling begins at the bottom of the mantle and moves upward very quickly. This is called a mantle plume: The hot material forms a mushroom-shaped head at the top of the plume, and the remaining hot material trails behind like a thin tail. The hot head of the plume can melt as it reaches shallow depths in the planet, and the resulting melt can erupt onto the surface as volcanoes. The Hawaiian island chain is thought to be an example of a mantle plume on Earth: As plate tectonics moves the Pacific ocean plate over the otherwise stationary plume, the volcanoes emerge through the plate in a line. Such a plume may have formed Tharsis on Mars. Special considerations must be made, though, to explain how a plume could rise far enough to melt under what had to have been a thick lithosphere, and more importantly, why there would be only one Tharsis on the planet. If conditions were right to make a giant plume, why not more than one? Scientists are generally displeased with theories that require what is called "special pleading," arguments on why this is an unusual case, or that rely on coincidence.

A second, intriguing theory on the formation of Tharsis involves Hellas, the huge impact crater that is almost perfectly exactly opposite Tharsis on the planet. What a strange coincidence that is, to have the two largest surface features exactly opposed! Scientists at several universities have created large numerical models to investigate how Mars's mantle might move in response to the impact. Though the impact can make strong mantle convection right underneath its center, at the present time there is no model that indicates a large volcano should form on the opposite side of the planet as a result of the impact.

A third theory for the formation of Tharsis emerges from models for early mantle evolution. Work by Marc Parmentier, Sarah Zaranek, and the author at Brown University indicates that if the planet was initially hot enough to have melted and formed a magma ocean, when it subsequently crystallized denser material would form near the surface. (See the previous section "Was There a Martian Magma Ocean?" on page 50.) This dense material would also contain much of the planet's radioactive elements, which do not fit into common mantle minerals and so are concentrated in the last liquids left as the planet crystallized. This dense, radiogenic material would then sink into the planet's interior. Sarah Zaranek, a scientist at Brown University, hypothesizes that this material would continue to heat up through radioactivity as it lay at depth in the planet. Eventually it would form a warm upwelling through the Martian mantle. If the material was warm enough, it might rise to depths shallow enough to melt and perhaps form the Tharsis rise. The warm conduit formed through the mantle by the rising material might persist over time, allowing continued activity on Tharsis.

Mars has four main shield volcanoes, Olympus Mons and the three shield volcanoes on Tharsis discussed above: Arsia, Pavonis, and Ascraeus Mons. Though volcanism is centered on and around Tharsis, Mars has evidence for volcanic activity over much of its surface. In addition to the shield volcanoes Mars has ancient, shallow volcanoes known as patera volcanoes. Apollinaris Patera, one of the largest, has a caldera that is alone 60 miles (100 km) in diameter. Alba Patera, to demonstrate how shallow these volcanoes are, is 1,000 miles (1,600 km) in diameter but only four miles (6 km) high.

Susan Sakimoto, a scientist at NASA's Goddard Space Flight Center, is comparing the shapes of shield volcanoes from Mars with those from the Snake River Plain in Idaho. Amazingly, Idaho has less accurate topographical data than parts of Mars has, and Sakimoto and her team have had to take handheld GPS units and remap the topography of the volcanoes they are examining in Idaho. They find both Mars and Idaho have three main types of shield volcanoes: low-profile shields, dome-shaped shields like cowboy hats, and steep-summit shields with jagged peaks in their centers. The topography of the shield volcanoes in Idaho is created by the viscosity of the magma that is erupting, that is, the steepest volcanic peaks are formed by highly viscous lava that hardens before it flows.

Sakimoto and her team suspect that the same mechanism may be at work on Mars.

Beyond the major shield and patera volcanoes, Mars has ample evidence for widespread, small-scale volcanic eruptions. Elemental maps of the surface from *Mars Odyssey* indicate a high potassium spot near the North Pole, and several high iron spots north of the dichotomy boundary. These may be volcanic centers. Many individual volcanic flows have been found on the planet through careful examination of photographs. One flow has been found that is 300 miles (480 km) long, with a width from three to 30 miles (5 to 50 km) and a height of 100 to 200 feet (30 to 100 m).

Both the Mars rovers have found abundant volcanic rocks at their sites. Each has measured the compositions of rocks using its instruments, and together they have greatly increased the information on rock compositions on Mars. Remote sensing can have difficulty inferring true compositions of surface rocks because, beyond the difficulty of deciphering mineralogy from one spectra, the rocks have been

The Mars rover Spirit *ground through the weathered rind of the rock dubbed Clovis to create a hole 1.8 inches (4.5 cm) in diameter where fresh rock could be examined.* (NASA/JPL/Cornell/USGS)

altered by weathering and covered with dust, obscuring their true internal compositions. The rovers carried tools that could grind through the outer coating on rocks, allowing other instruments to measure the internal composition. A rock dubbed Clovis is shown in the image on page 125 being ground by the rock abrasion tool to a depth of 0.35 inches (8.9 mm). The hole is 1.8 inches (4.5 cm) in diameter. Clovis is one of the softest rocks encountered in Mars because it contains mineral alterations that extend deeply from its surface, likely caused by acidic water as evidenced by its high levels of sulfur, chlorine, and bromine.

One small rock was struck by the *Opportunity* rover on landing, when the rover was still protected by a mass of large airbags. This 14-inch (35-cm) rock was named Bounce. After grinding the rock was available to the Thermal Emission Spectrometer, an instrument that measures the infrared radiation emitted by the rock from incident sunlight. The spectrum that Bounce gave off was the sum of the spectra of all its constituent minerals and so could be used in two ways: The spectrum could be compared to the spectra of rocks on Earth, whose bulk compositions are known, and the spectra of known minerals could be added up to see if they result in the spectrum of Bounce. Bounce's spectrum closely matched the spectra of Mars meteorites, though it is slightly enriched in silica. Bounce is therefore a basaltic rock similar to some of those that were knocked off Mars by impacts and traveled to Earth.

The *Mars Global Surveyor* orbiter also carried a Thermal Emission Spectrometer (TES), which detects infrared emissions from the planet at wavelengths from six to 50 micrometers while in orbit. The TES team, led by Phil Christensen of Arizona State University, identified two large regions on Mars that have distinctive spectral properties, though each is interpreted to be a dark igneous rock composition. In the southern highlands the dark igneous rocks appear to be basalts, similar to the magmas erupted from Hawaii on the Earth. In the northern lowlands the dark igneous rocks appear to be a type of lava higher in silica, known as andesite. On the Earth andesites are produced at subduction zones as the products of melting the mantle in combination with water. If these lavas really are andesites, then they may indicate something about the location of water and the role of plate tectonics on the early planet, or it may indicate that andesites can be formed in other ways unlike those on Earth.

Scientists had mixed reactions to the possibility of andesite on Mars. One question raised is how uniquely the spectra of Surface Type 2 matches andesite. Michael Wyatt and Harry Y. McSween, scientists at the University of Tennessee, and Timothy Grove from the Massachusetts Institute of Technology have taken another look at the Thermal Emission Spectrometer spectra by using a larger collection of aqueous alteration (weathering) products in the spectral mixing calculations. They show that weathered basalt also matches the spectral properties of Surface Type 2. Wyatt and McSween also note that Type 2 regions are generally confined to a large, low region that is the site of a purported ancient Martian ocean. They suggest that basalts like those in Surface Type 1 were altered in the ancient Martian sea. Independent data are needed to test the andesite versus altered basalt hypotheses.

Mars's extensive volcanism that has continued throughout its history is a problem for geoscientists. On Earth volcanism is driven largely by plate movements: Volcanoes form at convergent plate boundaries, boundaries where an oceanic plate is pressing against (converging on) a continental plate, and eventually sliding beneath it to form a subduction zone. Volcanoes on Earth also form at divergent plate boundaries, the midocean ridges along which new oceanic crust is formed and flows away. On Mars there is no plate tectonic movement, and if ever there was plate movement, it was a brief interval early in the planet's history. Mars's mantle must be convecting under its thick solid lithosphere and allowing mantle material to melt in upwelling mantle plumes. No one has yet devised a convincing model that explains continuous volcanic activity over large portions of the planet's surface for the length of the planet's history, while simultaneously building one single immense volcanic edifice (Tharsis).

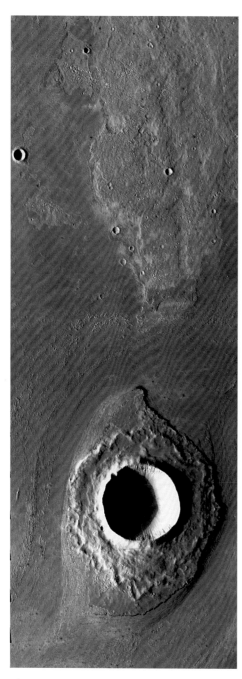

This rampart crater southwest of Athabasca Vallis has had material eroded from around its base, probably by ancient floods. (NASA/JPL/Arizona State University)

CRATERS

Mariner 4 in 1965 had so little computer memory and such slow transmission speeds (8.3 bits per second) that when it took an image of the Martian surface it would hold the data on tape and send the image back to Earth over 11 days time. Despite this agonizingly slow and fragile process, the *Mariner* photos were hugely influential in developing an understanding of the evolution of Mars. These photos were the first to show that the planet had craters; they could not be seen from Earth with the telescopes then available. As soon as the craters were seen, scientists knew that the surface of the planet was far older than had been thought. More recent processes like volcanism or flooding had not wiped away the marks of the violent, early solar system (though water floods have altered some craters on Mars).

If an impact is made by a small object, less than a few kilometers in diameter (but greater than a few meters, below which no crater is made), the resulting crater is shaped like a bowl, and referred to as a simple crater. Larger craters undergo more complicated rebounding during impact and end with circular rims, terraced inner wall slopes, well-developed ejecta deposits, and flat floors with a central peak or peak ring. These craters are called complex craters. A small simple crater with a rough, bouldery ejecta blanket surrounding it is shown here.

Several simple craters have been photographed by the Mars rovers. The largest is the crater Endurance in which *Opportunity* made several important discoveries about the history of water on Mars. The Endurance crater in about 525 feet (160 m) in diameter, but Fram

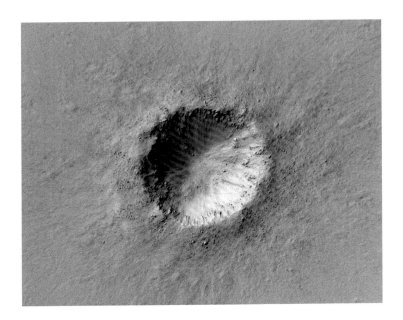

This small, simple crater is in the Arabia region of Mars. (NASA/JPL/Malin Space Science Systems)

(shown in the figure on page 130) is only about one-tenth that size.

Of the large craters, Hellas, Argyre, and Isidus are the youngest. They fell in a region of old crust just south of the crustal dichotomy that carried strong magnetic signatures from the time of its formation in the early years of Mars's evolution. Since the crust directly under the craters carries no magnetic signature, the impacts must have erased any magnetism that had been present in the crust before they impacted. Because no new magnetic field was recorded in the bottoms of the craters, the craters must have formed at a time after Mars's magnetic field had ended.

The scientist Pierre Rochette at the Université d'Aix-Marseille and his colleagues have investigated how giant impacts can erase magnetic fields in the rocks they strike. The heat of impact alone is insufficient, as can be shown by comparing the large area that is demagnetized by an impact to the

The rover team plotted the safest path into the football field–sized Endurance crater, eventually easing the rover down the slopes to investigate the rocks within. (NASA/JPL/ Cornell)

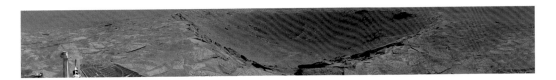

smaller area that is heated, according to models. They find that the mineral pyrrhotite is the likely phase to hold the magnetic field, and that it loses its magnetic field when pressurized to about 30,000 atmospheres (3 GPa). The scientists found during laboratory experiments that when the pressure is removed, if the planet were still creating a magnetic field, the field would be recorded in the minerals as they returned to normal. This finding supports the reasoning that a lack of magnetic field in the craters indicates that the planetary magnetic dynamo had ceased.

The oldest craters seem to be Aries and Daedalia, which have strong magnetic anomalies, and thus are thought to have formed while the Martian magnetic field was active. According to theories of pressure and magnetism, when the shock pressure of the impact was released, the minerals in the craters recorded the magnetic field of the planet at that time. These craters are easily large enough to create this effect, so the magnetic fields recorded in the crust under the craters indicates that Mars still had an active magnetic field at their time of formation.

There are three main minerals that retain magnetism from a planet's magnetic field: magnetite, hematite, and pyrrhotite. Hematite is hard to demagnetize with shock. Magnetite demagnetizes at more than 5 GPa, and pyrrhotite at about 3 GPa. Some of the magnetism around the crater therefore would have been removed by the pressure of the meteorite impact. Additional magnetism would be removed by the heat of impact. The question may then become, how is any magnetic field retained at the site of a giant impact? It is most likely to be the result of having an active planetary field to remagnetize the rocks while they are still hot or under pressure.

This panoramic image of the small crater Fram was taken by the Mars rover Opportunity. Fram *shows a few dust ripples in its boulder-free bottom.* (NASA/JPL)

The shape of craters on Mars also provides a piece of evidence for surface water in the Noachian. Many of Mars's craters have peculiarly shaped rims that are called ramparts. Rather than a typical rim structure, many Martian craters have a splash-formed rim, thought to indicate that the impactor struck either a water- or ice-rich surface layer. Some of these craters also have associated splash-shaped ejecta on the surrounding plains. Virtually all the craters in the northern hemisphere with diameters greater than about two miles (3 km) have ramparts. Because of the number of craters with ramparts, and the ability to date some of the craters using Mars's relative age scale, it may be shown that water was ubiquitous in the Hesperian and maybe even more recently.

Two rampart craters shown on the right have typical lobate ejecta blankets that were formed from fluidized ejecta, presumably from liquid water. The crater bottoms are smooth for the same reason. For comparison, a fresh, young Martian crater is shown on page 132, displaying a central peak at the bottom of the crater and long, dry ejecta rays (field of view is about 1.9 miles [3 km] across).

Craters provide a method for estimating the age of the surface they are on. On planets such as Mars, Venus, and Mercury, surfaces are "dated" using crater ages, since few or no rocks are available for dating using radionuclides. Based on information from the Moon, the heavily cratered terrestrial body about which scientists know the most, cratering rates were much higher in the distant past of the solar system. Then, presumably, there was far more material loose in the inner solar system to bombard the planets. As time progressed more and more of the material

The splash-shaped ejecta blankets from these rampart-rimmed craters are thought by some researchers to indicate that the impacts occurred in material containing water. (NASA/JPL/Arizona State University)

had impacted a planet, fallen into the Sun, or attained some sort of stable orbit, and so cratering rates declined. Scientists count craters, measure their diameters, and plot the results on graphs that show the number of craters versus their diameters. Planetary surfaces of different ages create parallel lines on these plots with older surfaces showing higher numbers of craters of all sizes. Through arguments about known or suspected ages of features on the planet these curves can be approximately related to absolute ages. Though cratering ages are fraught with uncertainty, they are often the best estimates that scientists have.

SURFACE FEATURES CREATED BY WIND AND WATER

Photogeologists spend much of their time trying to understand and model the shapes of surface features. Understanding surface features can help to determine compositions and textures of surface materials (for example, certain kinds of sand dunes form only from grains of certain sizes) and also to constrain weather and surface volatiles. One of the biggest questions

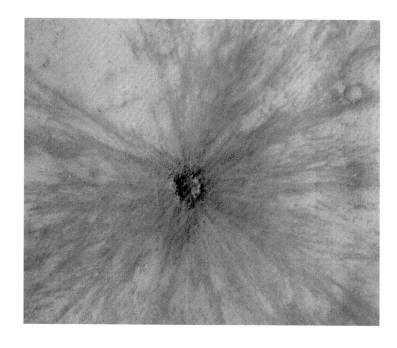

This fresh crater, by comparison with the rampart-rimmed craters in the previous figure, shows the starburst pattern of linear ejecta rays characteristic of impacts in dry material. (NASA/ JPL/Malin Space Science Systems)

about Mars concerns the evolution and current disposition of the water on the planet. The action of liquid surface water on the planet in the past will have left its mark in specific kinds of erosion: gullies, chasms, smooth lake bottoms, and terraced shorelines that lie at a single elevation. Though there is no liquid surface water now in any quantity or purity, the appearance and disappearance of streaks may indicate seeping surface water now. Ice in the soil certainly exists, since it has been detected by orbiters, and some evidence for ice and rock glaciers also exists on the surface today.

The giant Valles Marineris canyon system can be seen from space. (NASA/JPL)

Mariner originally made images of riverbeds and caused great excitement: These valleys are the proof that in times past, liquid water flowed on Mars's surface. The largest of the valley systems is Valles Marineris, a system of channels 2,800 miles (4,500 km) long with a maximum depth of four miles (7 km). The scale of these channels is immense, not only in length but in depth. On Earth the Valles Marineris would stretch across the entire United States. The southern Melas Chasma portion of the Valles Marineris displays light-colored, layered, sedimentary rocks in the image on page 134. The image is about 1.1 miles (1.8 km) across.

The river channels shown on page 135 were created by flowing water, not lava, as evidenced by their braiding back and forth, the lack of accompanying lava flows, and the lack of crusts of lava lining the edges of the river channel. Valles Marineris is thought to have been formed by floods of water in the ancient past of the planet, probably before 3.5 billion years ago. This image was taken from orbit by *Mars Odyssey.*

Sedimentary rocks crop out in Melas Chasma, while dunes fill the bottom of the canyon. (NASA/ JPL/Malin Space Science Systems)

The sinuous channels and blunt-ended box canyons are typical of landforms created by water. Though similarity between photos cannot alone prove that two landforms were created by the same process, detailed analysis of angles, cross sections, sediment patterns, and other specifics between Martian and Earth channels has convinced the scientific community that the channels on Mars were formed by running water. The *distributary* fan shown in the image on page 138 is about eight miles (13 km) wide. Only long-lived water river systems can create fans of this type: At the end of their drainage, the river water slows and drops its sediment load into this typical shape. On page 136, the Mississippi River delta is shown as a terrestrial comparison.

For a short time the question remained in the scientific community of whether the channels might have been formed by continuous and long-lived seepage of groundwater. Several points have proved that the channels were formed by collection of precipitation, as they are on Earth. Some channels start at topographic highs, proving that groundwater seepage could not have formed them. (Groundwater flows downhill before it emerges through a hillside onto the surface, so groundwater cannot make flow patterns on the tops of hills.) Careful analysis of the volumes of apparent ancient lakes in craters along with the volume of sediment carried into the crater by water flow shows that in some cases the amount of water exceeded the amount of sediment by a factor of 100 or more. Groundwater transport necessarily carries more sediment with it, so the high water volumes also prove that surface water was the source of the flow. The image in the figure on page 137 and others like it of multiple complex channels in the Warrego Valles region in the southern hemisphere are an example of evidence for liquid surface flow. When viewed at very high resolution (1.5 to 4.5 meters per pixel) with the *Mars Global Surveyor* images (see the figure on page 137), the Warrego valleys break down into a series of vaguely continuous troughs that have been covered and partially filled and then re-eroded to form a very rough-textured surface. None of the original valley floor or wall features are visible.

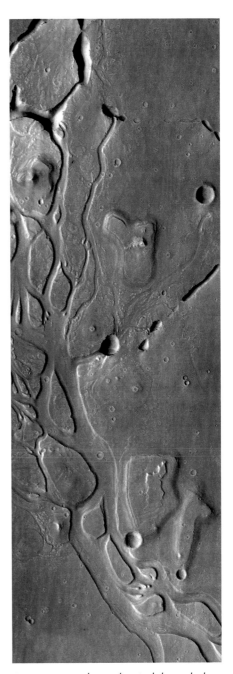

Ancient river channels wind through the Elysium Planitia, while collapsed lava tubes lie in the plains nearby. (NASA/ JPL/Arizona State University)

The Mississippi River delta showing its distributary fan formed by sediment dropped from the river water as its flow slows during the gradual descent to the ocean. (NASA/GSFC/ METI/ERSDAC/JAROS, and U.S./Japan ASTER Science Team)

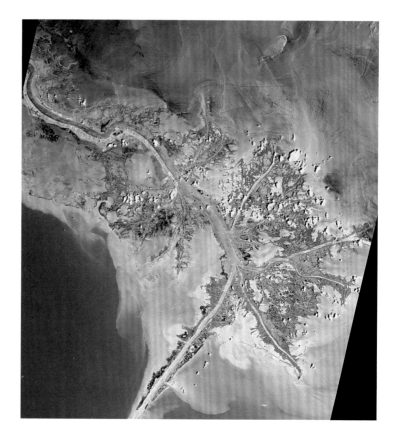

Areas on Mars, particularly those near the ends of outflow channels apparently caused by flowing water, form complex fractures and plates called chaos zones. In these settings the chaos is apparently formed by the rapid removal of subsurface water, either by draining or by sublimating water that had frozen into the soil. The chaos zone shown on page 140 appears in a crater near Elysium Planitia, far from an outflow channel, so its formation is more of a mystery.

The scientific community now almost universally agrees that Mars had flowing surface water caused by precipitation of rain or snow in the past. The question remains, though, whether or not Mars ever had enough standing surface water to constitute a sea. The channels now left carved and dry on Mars must inevitably have carried water to lakes, but were conditions right to allow great lakes to persist and grow into oceans? There is no clear consensus on the possible existence of liquid oceans in the Martian past.

A great example of photogeology is being carried out by a number of scientists, including Jeff Kargel, a scientist at the U.S. Geological Survey (USGS). (See the sidebar called "Jeff Kargel and the Search for Ice on Mars" on page 142.) Photogeologists have been studying peculiar shapes along the southern sides of mountains on the Tharsis rise. These surface

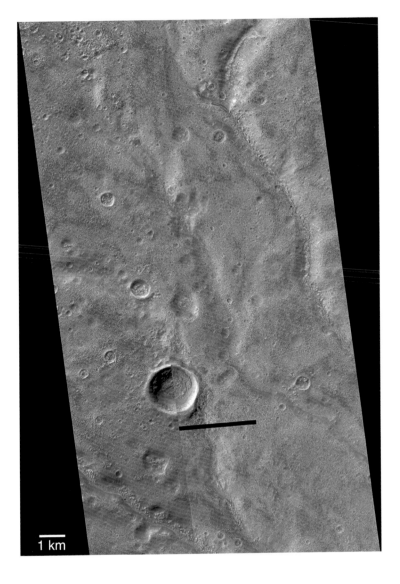

1 km

Multiple channels are visible in the Warrego Valles region of the Martian southern hemisphere: This high-resolution image shows their extreme state of erosion and weathering. (NASA/JPL/ Arizona State University)

This large distributary fan on Mars is evidence for long-term water flow: The fan measures eight miles (13 km) across. (NASA/JPL/Malin Space Science Systems)

shapes are smooth surfaces down the sides of the slopes, with curving deposits along their bottoms. These deposits resemble the terrestrial deposits left behind by ice glaciers: The glaciers carve out smooth valleys as they creep down slopes, and they deposit piles of rocks and soil where they end. On Mars, of course, there is no earth to pile, but more important, it is not possible for ice deposits to form on Mars near the equator; temperatures are too high.

Using the meticulous new measurements of Mars's orbital and spin behavior, other scientists have made computer models that incorporate the physical forces Mars experiences as it orbits the Sun and have calculated the likely evolution of Martian spin and orbital behavior that led to its condition today. Multiple runs of these models suggest that Mars is at the moment in an unusual state, with an obliquity (inclination of the equator to the orbit) of about 25 degrees, similar to the Earth's. The models suggest that Mars has more often had a much higher obliquity, and that about 5 million years ago, Mars's obliquity was 45 degrees. The models also indicate that Mars's eccentricity varied up to 0.12.

Mars's obliquity changes radically over time because it has no stabilizing moon; Earth's Moon prevents Earth from suffering a similar fate. The small variations in Earth's orbital

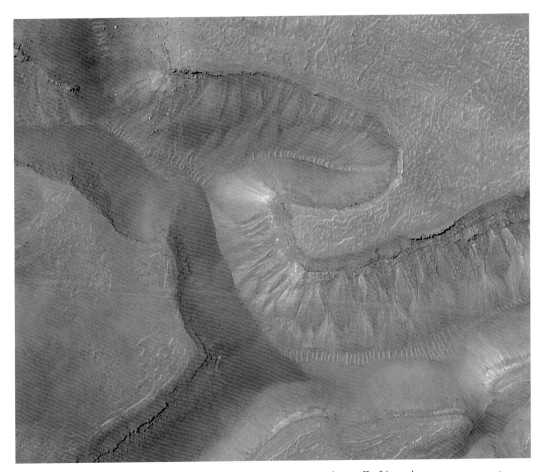

These gullies, proposed to have been formed by seepage and runoff of liquid water in recent times, are located on the south-facing wall of a trough in the Gorgonum Chaos region. Below the layer of seepage is found a dark, narrow channel that runs down the slope to an apron of debris. Sharp contrasts between dark and light areas are hard to maintain on Mars for very long periods of time because dust tends to coat surfaces and reduce brightness differences. There is no way to know how recent this activity was, but educated guesses center between a few to 10s of years, and it is entirely possible that the area shown in this image has water seeping out of the ground today. (NASA/JPL-CalTech/University of Arizona)

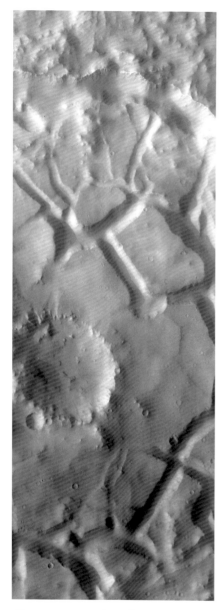

Chaos terrain such as this in a crater south of Elysium Planitia is thought to form from the sudden removal of subsurface water or from gradual frost wedging of water that sinks in and freezes. (NASA/JPL/Arizona State University)

parameters appear to create a long-term climate cycle called the Milankovitch cycle, but its severity is no comparison to what happens on Mars. Together, a more eccentric orbit and a high obliquity on Mars make for far more extreme seasons, both colder and hotter. The colder seasons are consistent with surface ice accumulations in some locations of up to about an inch (3 cm) per year, enough to create glaciers on the surface of Mars. These calculations also suggest that Mars had snowfall as recently as 5 million years ago! Though the most recent period in the model runs was about 5 million years ago, the models also indicate the Mars's obliquity changes on cycles of about 100,000 years, and so this period of very low obliquity is unusual. A new study using observations from NASA's *Mars Global Surveyor* and *Mars Odyssey* orbiters concluded that polar warming during periods of high obliquity caused mobilization of water vapor and dust into the atmosphere, cooling the planet's surface and causing the growth of surface ice and dust down to about 30 degrees latitude in both hemispheres, the equivalent of the southern United States or Saudi Arabia on Earth. An artist's image of Mars at the time of this ice age is featured on page 141.

On Mars today there are also flow features that look very much like glaciers, only they are entirely the color and texture of the surface rock and dust. These flow features occur at mid-latitudes, not near the icy Martian poles. They have been termed "lobate aprons," "lineated flows," and "furrow and ridge topography." The lobate aprons occur around local high points, and they look as if material had simply slumped away from the high point and formed a puddle with a lumpy edge around its bottom, something like the surface of maple syrup would look if it were poured over a strawberry and pooled around its bottom. Lineated flows are seen

Because of changes in Mars's obliquity, the planet goes through ice ages in periods on the order of several hundreds of thousands of years. This image is an artist's simulation of what Mars would look like during one of these ice ages. Recently discovered ice-field equatorial glaciers indicate that some ice ages encompassed areas closer to the equator. (NASA/JPL)

in sloping valleys: The valley seems to be filled with soil, but the soil has lines along it that parallel the valley walls. Furrow and ridge topography is much as it sounds, alternating furrows and ridges aligned with the direction of downhill flow.

All these features are similar to ice flow features on Earth. Lineated flows are seen in valley features, where the flow of ice causes it to form lines of ice and rocks parallel to the valley walls. Ice glaciers on Earth also make furrow and ridge topography. These Martian glaciers are called tropical cold-based glaciers, because they flow like glaciers but are found in midlatitudes and because they require a cold base layer of soil where ice can exist. Examination of these features and computer models of the assumed processes indicate that these cold-based glaciers flow at five to 10 centimeters per year. Many exist in the topographically high region around Tharsis Montes.

(continues on page 144)

JEFF KARGEL AND THE SEARCH FOR ICE ON MARS

Like many people now in science, as a child Jeff Kargel fell in love with nature. His current fields of expertise in the compositions and mineral content of icy moons and in Martian geomorphology (the study of the shape of the land) may be a far cry from tromping through fields and patches of forest in search of insects, but fascination with the natural world tends to expand beyond the boundaries of one discipline. His specific interest in geology began early, as well, with a passion for fossils and an intense years-long search to find a particular specimen called a trilobite. This quest led him to rip apart a retaining wall near his parents' garage in hopes of discovering one, but somehow the seriousness of his interest spared him too much punishment.

Kargel's grandmother was educated only through the sixth grade, but her obsession with space carried her to heights of self-education. She kept press clippings about every space launch, and she collected rocks she thought to be meteorites. Her love of her field was contagious. One of Kargel's most intense childhood memories is watching the live television broadcast from the Moon on Christmas Eve, 1968. Throughout the Apollo era his interest grew, to the point that his mother would give him notes to excuse him from school so he could watch daytime television broadcasts of space missions. Eventually this drive took him through college in aerospace engineering and then into planetary geology, culminating in a doctorate from the excellent program at the University of Arizona at Tucson, and his current position at the United States Geological Survey (USGS).

Kargel has become one of the world's experts on ice and aqueous processes at low temperatures. This specialty allows him to study such disparate topics as the tectonics and volcanism of the icy moons of the outer planets, the landforms of Mars that appear to have been formed by ice, and glaciers and permafrost regions here on Earth. He is the lead scientist of a 25-nation consortium, including 81 separate institutions, that studies glaciers on Earth from space.

Kargel says, "I just love glaciers, they are so beautiful and seductive and their dynamics and form just draw you in. I feel very strongly about what humanity is doing to our own planet in terms of global climate change, which can be seen clearly in changes in glaciers. The international collaboration is great: we should get along on a human level even if we don't see eye-to-eye politically or religiously."

Currently Kargel is also working on problems related to ice and glaciers on Mars. Mars has permafrost and debris-covered glaciers at mid-latitudes, the vast southern polar ice cap, and evidence for ancient continental ice sheets. He seeks to understand the relations of these ice deposits to climate change due to Mars's obliquity variations, and between these ice deposits and ancient oceans and outburst floods and volcanic-ground ice interactions. Most important, what are the relationships among water, ice, climate change, and life? Water is required for life, at least as scientists now understand it. There might have been life on Mars in the past, and there might be life there now. Studying Mars and determining whether life has or has never existed on the planet is critical to understanding of the solar system and ourselves, and mankind will almost certainly know in the next years or decades whether life has existed on Mars.

To study ice on Mars Kargel uses images from the Mars Orbital Camera (MOC) and the Thermal Emission Imaging System (THEMIS) on the Mars Global Surveyor mission, along with some data from the Viking orbiter and topography from the Mars Orbital Laser Altimeter. Both the MOC and THEMIS data allow Kargel to look at the three-dimensional form of the surface to understand processes, conditions at the time the forms were produced, and sequences of past events, a geological field called photogeology. A lot of photogeology is an art, Kargel says. How, for example, does one know the surface form seen in the image is a debris-covered glacier? Kargel looks for textural features, like linear corrugations, pits, and knobs, and the patterns they form: Do corrugations lie across or down the slope? On Earth, the ridges of rocks and soil carried along and molded by a glacier lie down the slope, along the sides and under the glacier (these are called medial and lateral moraines). Kargel therefore looks for places on Mars where lines of mounds or ridges lie down the slope. He also looks for places where ridges of higher topography trace uphill to a knob or a junction between two valleys, and for curving arcs of boulders along the front where the potential glacier ends, both features found in glaciers on Earth. Glaciers on Earth also form eskers underneath them, sinuous ridges of rocks and gravel left by streams flowing beneath the glacier, and so similar features are searched for on Mars.

Kargel looks for places where two or more of these features occur together, making the possibility of their formation by a glacier more likely. There are no esker-like features in the mid-latitudes on Mars, but they exist on a magnificent scale in the southern Argyre basin; some sets of braided ridges there are 190 miles (300 km) long. When Kargel interpreted these braided ridges as glacial eskers back in the early 1990s there

(continues)

(continued)

was a furor in the scientific community, but now that so many more features created by ice have been suggested by other photogeologists, the idea of giant esker systems is more palatable.

Now that these and other glacial features have been identified with some level of certainty, the essential question becomes determining sequences of events as well as their relative ages. The newest missions to Mars have provided tremendous quantities of data on crater counts and have brought some surprises in terms of crater density. Rarely are there as many small craters as expected based on the numbers of large craters seen in images. Broad areas of the planet have no small craters that can be discriminated in the Mars Orbital Camera images. The lack of small craters may indicate that surface features are constantly being erased by surface processes. There are Aeolian (wind-driven) processes operating constantly, but over a time frame of few hundred thousand years Mars also has active ice-driven processes, including permafrost and freeze-thaw processes that churn soil, creep of icy masses down slopes, local outbursts of briny groundwater of some sort, and possibly melting of surface ice deposits causing debris flows. None of these occur at the rates of surface processes on Earth, but they occur fast enough to erase small surface features over sections of the planet.

As new and more detailed data on the surface of Mars continues to be provided by missions, Jess Kargel and other Mars photogeologists will continue to identify ice-driven surface processes on Mars. The great new challenge is to determine more clearly when these processes have been most active. The overall water budget on Mars probably has been partly liquid and partly frozen for most of the history of the planet, but the timing of the warmest periods of the planet's history is still not determined, nor is the abundance of liquid water on the planet today.

(continued from page 141)

On Mars today, though, surface temperatures make ice glaciers impossible, and so it was thought that these features were made of the same material that the surrounding surface is, that is, rocks and dust. The interpretation given to these features was that the pore spaces of the soils at depth are filled with water ice, or possibly frozen carbon dioxide. The ice in the interstices of the soil particles was thought to allow them to flow over time more readily than the soil would itself. In effect, these were thought to be soil glaciers with ice particles, where glaciers on Earth are ice glaciers with a little soil and rocks mixed in.

New data from the *Mars Reconnaissance Orbiter*'s SHIRAZ ground-penetrating radar shows that many of the rocky flows are actually filled with ice today, even in the current climate. This ice is a big surprise, since the climate at Mars's equator would not allow the formation of ice today. Its existence shows that there was very recent climate change on Mars, so recent that ice that could form in the past climate has survived into the present climate. This climate change may have been caused by violent changes in the inclination of Mars's axis of rotation, which are thought to happen regularly because Mars's moons are not large enough to stabilize the planet's orbit. The flows in this region seem to have been active for about 100,000 years, based on calculations of how much material has flowed, and this is also consistent with the calculations that show that Mars's obliquity changes on about 100,000-year cycles.

Mars had flowing surface water in channels in the distant past and now has a few cold-based rock glaciers, but its surface is now defined by windblown sediments. Mars has a number of well-formed dune fields, with dunes shaped like those on Earth. The dunes are mainly in the north polar sand sea, in the high latitudes of the smooth northern plains. Others are in dune fields between craters at 30 to 60 degrees south latitude. Many of these dunes are the type called barchan dunes (see figure at right), great crescent-shaped dunes with the horns of the crescent pointing downwind. From studies on the Earth it is apparent that the more curved the crescent, the higher the average wind speed. If the two horns of the crescent are different lengths or have different curvatures, then the wind is variable. By

Dark barchan dunes lie in the bottom of Arkhangelsky crater, just to the northeast of the giant Argyre impact basin in the southern hemisphere of Mars. (NASA/JPL/Arizona State University)

A transverse dune field in Mars's northern plains consists of dark material thought to originate in the northern polar-layered deposits. (NASA/JPL/Arizona State University)

studying the shapes of Martian dunes in this way, some information can be obtained about weather on Mars.

The length of barchan dunes on Mars is 500 to 1,000 feet (150 to 300 m), while the width is 1,000 to 1,400 feet (300 to 430 m), similar to the dunes on Earth. On Earth, the height of the dunes is about 10 percent of their length, and the same ratio applies to the dunes in the northern plains on Mars. In the southern inter-crater fields, the dunes have different height to length ratios. This may be due to an ice content in the northern plains: The northern dunes are in an area where high amounts of hydrogen have been detected, probably the signature of ice in the regolith layer. Ice in dunes on Earth has been shown to stabilize them and allow them to grow taller, and perhaps the same effect is happening on Mars.

Mars also has the other common types of sand dunes found on Earth, transverse dunes (lying across the wind direction) and longitudinal dunes (lying in the wind direction). Transverse (see figure at left) and longitudinal dunes form where there is a larger sand supply than would allow the formation of barchan dunes.

A second example of transverse dunes (see figure on page 147) was taken by the *Mars Global Surveyor* Mars Orbiter Camera and shows small windblown transverse dunes, perhaps better called large ripples because of their size. Their orientations indicate that the responsible winds came from either the northwest (upper left) or southeast (lower right). The more complex ripple patterns within the two large craters result from local changes in wind direction caused by the steep sides of the craters.

Other dunes on Mars resemble giant ripples, just like those formed on Earth at the bottom of large, fast-flowing rivers, or where floods have occurred. There are areas on Mars where cliff-like barriers mark the edges of what appear to be great flood channels. The bottoms of these channels are lined with giant ripples. They well may be places where huge liquid water floods once raced over the surface of Mars. In still other places, the dusty, sandy surface of the planet has formed into dunes with shapes unlike any on Earth, for example, looking like giant apostrophes. Without an analog on Earth, there is so far no theory of how these dunes are formed.

The spectral and meteorite information show that, beneath the regolith, the surface of Mars consists largely of igneous rocks, coated everywhere with oxidized products of weathering rocks. When igneous rocks are made wet with water, or brought into contact with other reactive, oxygen-bearing compounds such as peroxide (which was detected by *Viking*), their constituent minerals break down into other minerals, most often clays when this happens on Earth. The iron in

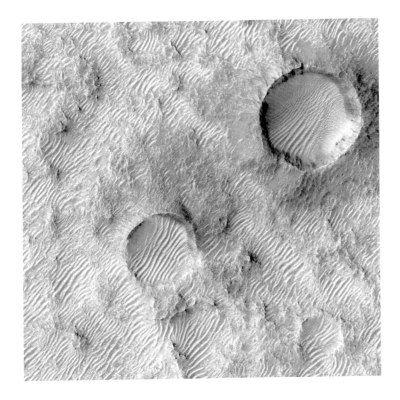

A rugged surface southwest of Huygens Basin is covered by large windblown ripples: The picture covers an area about 1.9 miles (3 km) across. (NASA/JPL/ Arizona State University)

these minerals is more oxidized (there are more oxygen atoms attached to each iron atom) than the iron in the original igneous minerals, and oxidized iron is colored a deep red. This is the source of the red dust that covers the planet and tinges the atmosphere. The sand dunes, however, need to be created out of larger grains. Tiny grains of dust cannot stack up to make the familiar shapes of sand dunes. On Earth, quartz grains make up the majority of sand dunes. Quartz does not weather into clay mineral flakes or into dust-sized particles; it remains as sand-sized material. On Mars there is no evidence for the formation of *granite* or other highly silica-rich rocks (quartz is pure silica). Earth, in fact, is the only planet in the solar system thought to have granites. (Scientists are discussing the possibility of missions to Mars specifically to search for granites, thinking that there may have been enough water for them to form.) On Mars sand dunes are dark-colored and thought to consist of grains of volcanic rock. The constant windblown sand also acts as a significant erosive force on Mars. The wind literally sandblasts the rock outcrops, and eventually they form distinctive shapes called yardangs. Yardangs are boat-shaped outcroppings that have been worn away by sand carried in the wind until they form a sharp point on their upwind ends, and their sides are worn into smooth curves. The image of yardangs shown in the figure on page 150 was taken by the *Mars Global Surveyor* camera and shows a region about 1.9 miles (3 km) square.

Though all evidence points to the importance of windblown sediment as the force that makes and erodes surface features on Mars today, the thin atmosphere lacks the strength to carry sediments on a regular basis. One of the Mars Exploration Rovers was required to stay in one place for 18 days during a period when the Sun blocked transmissions, and its camera photographed the same small patch of ground before and after its 18-day hiatus. Over the 18 days three or four of the grains of sand on the patch of ground rolled about their own length. The majority of grains did not move at all. The fierce dust storms created by the vast amounts of volatiles being released from the southern polar cap in southern spring are perhaps

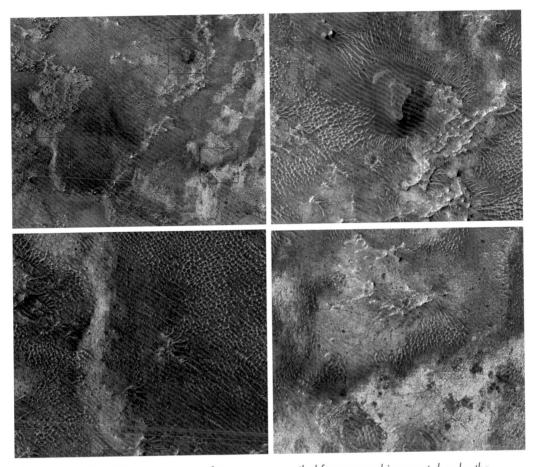

This image of the Nili Fossae region of Mars was compiled from several images taken by the CRISM spectrometer and the HiRISE camera on the Mars Reconnaissance Orbiter. The two major minerals detected by the instrument are olivine, a mineral characteristic of primitive igneous rocks and shown here in red, and clay, a mineral made by water interacting with rock, shown here in blue. The area being studied is at the upper left and is about three miles (5 km) across. The three boxes outlined in blue are enlarged to show how the different minerals in the scene match with different landforms. In the image at the upper right, the small mesa at the center of the image is a remnant of an overlying rock layer that was eroded away elsewhere. The greenish clay areas at the base of the hill were exposed by erosion of the overlying rock. The images at the upper right and lower left both show that the reddish-toned olivine occurs as sand dunes on top of the greenish clay deposits. The image at the lower right shows details of the clay-rich rocks, including that they are extensively fractured into small, polygonal blocks just a few yards (meters) in size. These images show that the clay-rich rocks are the oldest at the site, that they are exposed where overlying rock has been eroded away, and that the olivine is not part of the clay-rich rock, but occurs in sand dunes blowing across the clay. (NASA/JPL)

the major driving force for sediment transport on the planet today; strong and unusual weather is required to lift grains and drive them into dunes.

In 2004 the Mars Exploration Rovers *Opportunity* and *Spirit* sent back data that, to a large portion of the scientific community, proves unequivocally that there was flowing water on the surface of Mars in the past. Previous evidence has come from landforms that appear to have been formed by rivers and floods, from the smoothed northern plains, and from inferences about the volume of frozen water on Mars today and the likely temperatures on the planet in the past. The evidence from the rovers came first from a rock outcrop near which *Opportunity* landed. The X-ray spectrometer on the rover found strong emission lines that indicate that the rock has a large percentage of sulfur, sulfate salt, and iron sulfur hydrate. These minerals only form in water that is evaporating gradually, condensing trace elements in its remnants until they saturate the water and precipitate as miner-

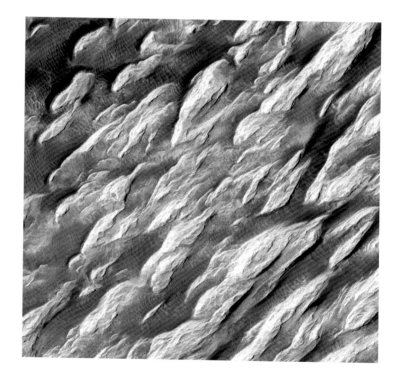

The bright rocks in Pollack Crater have been known since the Mariner 9 mission in 1972. Wind carrying dust has sculpted the light-toned material into ridges and troughs known as yardangs. (NASA/JPL/ Malin Space Science Systems)

als. Looking at the soil near the rock, the rover sent photos of a very fine dust like cocoa powder, studded with little spherical rocks, which the scientists likened to blueberries (an example is shown in the figure above). These small round rocks are made of hematite and are most likely to have formed out of liquid water.

Rounded, hard mineral agglomerations like these form in gaps in existing rocks when liquid water runs through the rock, and they are known as concretions. On closer inspection of the rock itself, the hematite concretions could be seen in the rock where they formed, between and among bedding planes (layers of rocks) that were created as sediment was added to the growing rock. These results were first presented by Steve Squyers, the Mars Exploration Rover principal mission scientist, at the Lunar and Planetary Science Conference in Houston, Texas, in March 2004. The largest hall at the conference was filled to overflowing, and scientists were standing out in the hallway, peering through the forest of people packed in the hall, trying to see the slides of Martian rocks through any tiny gap in the crowd.

The bedding planes were hailed by many as the final proof of liquid surface water. When a body of water gathers sedi-

These hematite concretions, dubbed "blueberries" by the Mars Exploration Rover team, are believed to have formed in water-rich sediments. (NASA/JPL/Cornell/USGS)

ment on its bottom, the movement of the water above the sediment molds the grains into ripples. Inside the ripples are layers of sediment that have been pushed by water movement into the ripple shape, and these layers are at an angle to the bottom of the body of water. Sediment layers that are parallel to each other but at an angle to the principal bedding plane (in this case, the plane of the surface of the planet) are called crossbeds. (See the sidebar "Wesley Andres Watters: Working on the Mars Exploration Rover Science Team" on page 153.) Over long periods of time, the sediments can harden into rock. The process of hardening into rock often allows the crossbeds to remain visible in the rock, showing the original process that formed the rock long ago. While the sulfur minerals and the "blueberry" concretions may have been formed by seeping liquid groundwater, the crossbeds were most likely formed by moving liquid water that lay in large pools on the surface of Mars in the distant past. NASA called in outside specialists in sedimentary rocks as a peer review panel before they

Images taken by the Mars rover Opportunity show the wide variety of Martian soil samples found near Eagle crater. Each image is slightly less than one inch (2 cm) square. (NASA/JPL/Cornell/USGS)

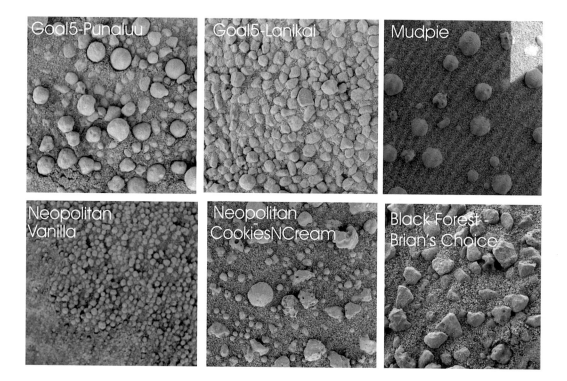

WESLEY ANDRES WATTERS: WORKING ON THE MARS EXPLORATION ROVER SCIENCE TEAM

Wes Watters was a graduate student working toward his Ph.D. at the Massachusetts Institute of Technology when he received a research position that would be envied by many senior scientists: He began work at the Jet Propulsion Laboratory (JPL) as a part of the day-to-day team of scientists running the Mars rover *Opportunity* (MER B). Watters was one of about 15 graduate students who cycled in and out of the JPL offices as their university coursework and other commitments allowed, the privileged students of the senior scientists on the rover teams. In many ways he had perfectly prepared himself for this task; as a child he voraciously read both paleontology and planetary science and as an undergraduate in college he studied both physics and math. His intellectual appetite drove him to work on more than one problem at a time, a source of distraction producing a lack of results in some scientists. In Watters's case, his wide-ranging interests paid off. By studying Mars, he could practice planetary geology, and his math and physics background qualified him well to work on the physical and practical aspects of the mission. Driven also by his interest in early life, he continued to work on Earth-based projects, including a study of ancient terrestrial fossils.

When morning came for the Martian rover *Opportunity*, Watters arrived at the MER B offices in the JPL flight operations building. He would join a meeting in which scientists plan the sequences (instructions for the spacecraft) for that Martian day. Each work group (groups include the soil, geology, long-term planning, and instrument team groups) reported their recent results, and each group had suggestions for the rover operations of the day. Watters participated as a member and, as the mission lengthened into months, as a leader of the long-term planning group, responsible for developing strategies of rover exploration on a daily and weekly basis. This entailed building a team consensus about what activities the rover would undertake: driving to interesting targets, deploying the instrument arm, taking remote observations using the rover cameras. The day's sequences were agreed upon and then sent to the engineers, who sent it to the spacecraft. Part of Watters's work also involved leading an effort to collect data for building a high-resolution three-dimensional computer model of the stadium-sized Endurance crater at the *Opportunity* landing site.

"I remember especially when we arrived at our deepest position in the crater and it was too dangerous for the rover to go further," recalls Watters. "There, we found things

(continues)

(continued)

that we hadn't seen before: unusual fracture patterns, cavernously weathered rocks, and a new kind of concretion weathering out from nearby rocks."

The rest of Watters's day was spent at his computer workstation writing computer code. Many of his programs concerned the three-dimensional terrain models generated from images taken by the rover's Pancam instrument, the panoramic camera with its 13 color filters. Watters was making computer models and maps of interesting features at the rovers' landing sites. One specific project he took on was mapping the "blueberry" concretions to measure their spatial distribution as well as how their distribution in the soil compared to their density in the rock outcrops. Watters also mapped the stratigraphy of the sedimentary rocks; that is, he measured the orientations of the bedding planes of sedimentary rocks and related one outcrop to another when their connection could not be seen because of soil cover. In short, Watters did with remote data exactly what a terrestrial geologist would be able to do in the field. Watters had been concerned about being forced into too narrow a specialization when becoming a scientist, and he worked against this common tendency in academic science. In addition to his work on the Mars rovers, he worked on computer models of the growth and shape of structures formed by early multicellular organisms and communities of unicellular organisms found in the fossil record. He also studied the formation of impact craters and the long-term consequences of large impacts upon the evolution of planetary surfaces and interiors. Watters successfully finished his Ph.D. at MIT in 2008 and moved on to a postdoctoral position working with Steven Squyres, principal scientist of the Mars rovers, and he is currently working again at the JPL.

announced these results; the finding is so important that no chance for mistake could be allowed. (Despite the care in data analysis by the rover team, some scientists still maintain that these same crossbeds could have been formed by wind-driven sediments.)

A variety of densities, shapes, and patterns of "blueberries" have been found and named (mainly after ice cream flavors!) by the rover team members (see figure on page 152). Continued analysis may help understand the surface and subsurface conditions that caused their varying formation.

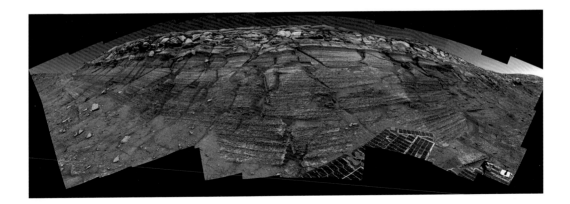

Though both rovers have found many examples of volcanic basalts at their landing sites and during their subsequent travels, each has also detected large volumes of sedimentary rock, like that from which the spherical "blueberry" concretions came. Impacts are a good way to see cross sections of bedrock on Mars, just as highway road cuts reveal bedrock on Earth. Burns Cliff (shown in the photograph above) is part of the side of Endurance crater. The rover *Opportunity* investigated the large crossbeds clearly visible in the cliff at the side of Endurance Crater. (This cliff was named in honor of Roger Burns, the late professor of geosciences at the Massachusetts Institute of Technology, with whom the author was privileged to study mineralogy.) A magnified view of crossbeds taken by the rover *Opportunity* is shown in the image on page 156.

In 1987, long before the detailed mineralogical information was sent to Earth from recent missions, Roger Burns modeled and described a young Martian surface environment in which volcanic activity poured sulfur into the atmosphere. The sulfur reacted to form sulfuric acid, which would have dissolved into any available water and corroded rocks. He suggested that this environment would produce a variety of sulfate and salt minerals, including a distinctive iron sulfate called jarosite. Although the scientific community doubted it could exist when Burns first suggested it, jarosite has now been identified on Mars, and all the information from both

Several panoramic images taken by the Opportunity rover were combined to make this image of Burns Cliff, part of the inner wall of Endurance crater on Mars. Burns Cliff is named in honor of Roger Burns, a mineralogist on the faculty of MIT who first suggested jarosite might be an important mineral on Mars. (NASA/JPL/Cornell)

Sol 41 The Upper Dells

0   1   2

cm

The bedding planes in this rock, called "Upper Dells," are curved and truncated against each other. This cross-bedded pattern is typical of rocks formed by sediments moved as ripples in flowing water. (NASA/JPL/Cornell/USGS)

orbiters and surface missions indicates that Mars was, and perhaps is, a highly acidic environment.

WATER ON MARS TODAY

Water flowed freely on the surface of Mars in the Noachian according to many planetary scientists. The large number of valley networks seen clearly on the Martian surface are widely believed to have been formed by flowing water; no other process has been conceived that can create landforms like those. There is no flowing surface water now.

An astonishing new discovery has been made that a large quantity of water ice resides in the pore spaces in the Martian regolith. Pore space is simply the amount of empty space between solid, rocky grains, and it is here that frozen water forms. The pore space might have been produced by condensation of carbon dioxide frost in the winter. The condensing CO_2 would have contained tiny grains of dust. When the CO_2 evaporated in the spring, it would leave behind a fluffy deposit of dust. Over time this fluffy, highly porous deposit might occupy the upper meter of the regolith, providing a medium in which atmospheric water could condense as ice. The Mars Odyssey mission produced the data showing there is a significant amount of water in the near-surface layers of Mars, and that these water deposits differ in the northern and southern hemispheres. These exciting results indicate that there is a large amount of frozen water in a thick soil layer, sometimes beneath a thin dry surface layer.

Detecting water frozen in the subsurface requires the use of the Gamma Ray Spectrometer. This instrument measures gamma rays from the *Odyssey*'s position in orbit. High energy cosmic rays interact with material on Mars's surface, and the materials on the surface give off gamma rays and neutrons. The speed of the neutrons is determined by what the cosmic ray initially hit: The *Odyssey* detectors can tell the difference between H, O, and C atoms, and thus recognize water within a meter or so of the Martian surface.

The *Mars Odyssey* instruments have detected large amounts of hydrogen within about a yard of the surface of Mars, with more ice at the poles and less at the equator. Hydrogen indicates the presence of water. Mars is cold enough, though, that

A thin layer of water frost is visible on the ground around NASA's Phoenix Mars Lander in this image taken on August 14, 2008, the 79th Martian day after landing. The frost begins to disappear shortly after 6:00 A.M. as the Sun rises on the Phoenix landing site. (NASA/JPL/ University of Arizona/ Texas A&M University)

the ground and any water in it is frozen to a depth of several kilometers. Ice is stable today in the top meter of Mars's regolith from the poles to 50 or 60 degrees latitude.

The *Phoenix* lander touched down at 68.22°N, 234.25°E, in the northern plains of Mars, on May 25, 2008. Its goal in landing in the remote northern plains was to investigate the frozen water there. It landed on the flanks of a shield volcano called Alba Patera on soil that likely mainly consists of a volcanic ash. This is both the farthest north and the youngest surface that any lander has reached on Mars. The lander scraped at icy soil that it uncovered during descent and found that the shards of soil sublimed away into gas, a good indication that the soil was largely made of ices. The robotic arm eventually succeeded in delivering a soil sample to the lander's measurement instruments, which heated the soil sample. The amount of energy needed to heat the soil indicates that the ice content is water ice. The lander was able to confirm in this way that the soil under its feet was rich in water ice from about one to six inches (three to 16 cm) depth.

The ground around the *Phoenix* lander shows an interesting pattern of cracks and bulges called frost polygons. These can be seen on Earth in Arctic regions where there are many freeze-thaw cycles during the year: In regions where there is permafrost winter cooling causes contraction (cool material is more dense than warm material in general). Now they have been identified on Mars to latitudes as low as 30 degrees north and south. A very few polygons are seen even closer to the equator.

The soil contraction from freezing causes some parts of the ground to fracture. Fractures typically form a honeycomb pattern across the ground, since this is the most efficient way to relieve the stress of contraction. During warm periods a little liquid water will sink into the fractures and, when it freezes, it expands. This is one of water's most surprising and unusual properties: It expands when it freezes, rather than contracting, like almost all other materials do.

The expanding ice in the fractures spreads the fractures wider and causes some bulging in the middle of the honeycomb pattern from pressure on all sides. Over long periods of

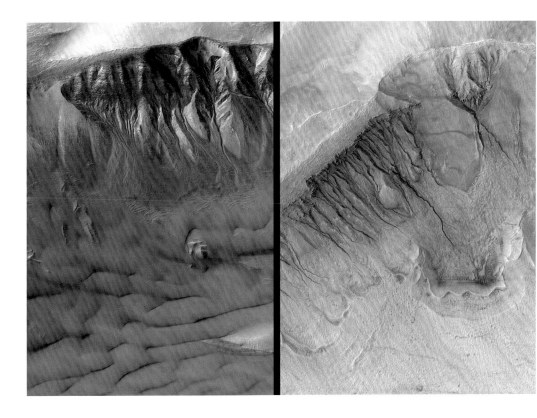

time the cracks get wider and wider, and the central bulges rise higher. Often in terrestrial Arctic areas the central bulges are filled with large stones through the ice wedging action of millennia, and the cracks contain only small stones. Mars, in comparison, has permafrost over a far larger fraction of its surface. The polygons seen from the lander vary from about 3.5 to 22 yards (three to 20 m) in diameter.

Images from *Mars Global Surveyor* show dark streaks appear on one side of certain valleys in the Martian spring, mainly at latitudes near the equator. Some early analyses suggested that these streaks are probably created by springs of subsurface water leaking through to the surface along steep slopes, but consensus in the scientific community is now that the avalanches are dry dust. They may still be triggered by sublimation of a small amount of ice in soil pore spaces, but the avalanches are intriguing and puzzling features. In every new image from Mars new dust streaks can be identified. Dust

These two images show dark streaks of dust in the walls of craters in Sirenum Terra. They were first thought to be wet but now are widely accepted to be dry dust streaks. The image on the left shows patches of wintertime frost on the crater wall. (NASA/JPL/ Malin Space Science Systems)

streaks seem to be forming at a rate of about one every 10 minutes, and old dust streaks are not fading away. Oded Aharonson, a scientist at the California Institute of Technology, and his research team are studying the locations and effects of frozen water in soil pore spaces from images, with computer modeling, and with experiments. Aharonson finds that ice in the near subsurface may be stable even at low latitudes where the surface is rough enough to form locations with little direct sunlight.

Unlike dark dust streaks, ancient gullies were thought for years to have been formed by flowing surface water. They are probably billions of years old, while the dark dust streaks are occurring in a short-term cycle of perhaps hundreds or thousands of years length, peaking at the present time. The ancient gullies are thought to have formed during the ancient period when flowing water formed channel networks and distributary fans. Though these ancient gullies were likely formed by flowing water, more recent gullies are now known to be only debris flows, with no water involved, according to new data from the *Mars Reconnaissance Orbiter*. This is one of the few cases when new data brought news that did not involve water.

The vast majority of water on Mars today resides frozen into pore spaces in the soil. The volume of this water is not well constrained, but estimates of the amount of water on Mars held in these soil reservoirs indicate a quantity equivalent to a global ocean layer a few hundred yards thick. The *Odyssey* results cannot address how much water could be present beneath the one-meter depth in which the gamma rays and neutrons are produced, so there may be still more water deeper in the planet.

At least two missions (the Mars Exploration rovers and the Mars Express orbiter) have detected minerals on the surface that have to be formed in the presence of water. These minerals are salts, sulfates (minerals containing the SO_4 molecule), and chlorates (minerals containing the ClO_3 molecule), including the minerals gypsum, anhydrite, antarcticite, jarosite, bloedite, and tachyhydrite, many of which are rare and obscure on Earth. While all these minerals require liquid water to form, they are also liable to dissolve into any liquid water that comes into contact with them. The existence of abundant salts and sulfates on the surface means that any liquid water is unlikely to be pure, because it will contain some percentage of salts and sulfur compounds. Jeff Kargel, a scientist at the USGS (see sidebar "Jeff Kargel and the Search for Ice on Mars" on page 142), has made models of water and salt systems at low temperatures that indicate there may be liquid water on the Martian surface at present. This liquid would have to be a very concentrated brine to be stable at Martian temperatures; pure water would be hard frozen, but brines have far higher freezing temperatures the high salt or even acid contents of these liquids lead Kargel to call the substance "icky water"). Chloride brines have been suggested, with 10 times the salinity of Earth oceans. The brines would dehydrate and form surface

Gullies in a crater wall in Noachis Terra are another possible example of erosion by flowing water in the distant past. (NASA/JPL/Malin Space Science Systems)

crusts. They may be erupted onto the surface from subsurface wells during particular seasons or days of warmth.

Even more stable than chloride brines are sulfuric acid brines. Three compositions of sulfuric acid brines are stable as liquids in specific low- and mid-latitude regions of Mars depending on their specific temperature and climate compositions. The least acidic of the three brines contains 35 percent acid and 65 percent water, and the most acidic contains 7 percent water and 93 percent sulfuric acid. Jarosite, one of the iron-rich sulfate minerals that has been found on Mars, requires a sulfuric acid brine to form (though not at the concentrations predicted), and so there is direct evidence for the existence of sulfuric acid brine.

Together the streaks caused by water seepage and the modeling that indicates acid-contaminated water may be stable at Martian surface conditions indicate that Mars likely has liquid water today both at depth beneath the cryosphere and perhaps nearer the surface in warm regions. The persistence of liquid water on Mars to the present day makes the search for life more hopeful yet.

Life on Mars?

Since the earliest days of Martian investigation with telescopes mankind has been searching for evidence of life on this sister planet. Once the initial sightings of green ("vegetation") patches had been debunked, and Percival Lowell's supposed canals built by intelligent life were no longer visible when better telescopes were used, hopes for civilization or even multicellular organisms were largely abandoned. Though Mars does not have canals, or any other evidence for intelligent life, many scientists now believe that it once had liquid water oceans, a thicker atmosphere, and a warmer surface, and so it is still a promising location for the development of life, particularly in its warmer past.

The meteorite ALH 84001 is famous for being the center of debate over the possibility of life on Mars. It is the oldest of the SNC meteorites (meteorites from Mars are collectively called the SNC meteorites, after the first three identified: Shergotty, Nakhla, and Chassigny) by far, having formed at about 4.5 billion years ago as shown by Sm-Nd isotopic dating. It was ejected from Mars about 15 million years ago and landed in the Allen Hills in Antarctica about 11 million years ago. It was discovered in 1984 but not recognized as Martian until February 1994. The first scientific paper suggesting the possibility of life as evidenced from ALH 84001 was published in August 1996. Since then well over one hundred papers have been published on this one rock.

ALH 84001 is an igneous rock, consisting primarily of the silicate minerals orthopyroxene, olivine, and *clinopyroxene,* with a number of additional minor phases. Voids between the igneous silicate crystals in ALH 84001 contain a range of carbonate minerals including calcite, magnesite, ankerite, and siderite, which in turn contain small crystals of magnetite and pyrrhotite. While the silicate minerals date to 4.5 billion years ago, the carbonate filling dates to between 3.9 and 4.1 billion years ago. The carbonate minerals are part of a secondary event, long after the rock initially formed. Carbonate minerals are all based on the CO_3 molecule, which unequivocally indicate that it formed in water (carbonate minerals of these types on Earth always form in water). Based on experience here on Earth, it seems that water is necessary for life, or at least the kind of life familiar on Earth, and so this proof of liquid water on Mars at some time in the past is very encouraging for those hoping to find life, or more accurately, signs of life in the past, on the planet.

The possible evidence for life is found in microscopic-scale textural features, crystal forms, carbon-based molecules, and oxygen isotope ratios, all found in carbonate filling in cracks in the meteorite. The carbonate minerals make up only about one percent of the rock, which in its entirety weighs only 4.2 pounds (1.9 kg). All of the evidence is microscopic, and so finding and proving the evidence for life exists and in fact came from Mars and not from later processing has become a kind of industry unto itself in the scientific community, complete with its own sessions at scientific conferences and adamant arguments for both sides held in person, at conferences, and in the scientific press.

The most famous possible evidences for life are lumpy, elongated microscopic imprints in the carbonate minerals that are proposed to be microfossils left by bacteria. These forms do look very much like terrestrial bacteria, but it has become evident from the violent arguments at scientific meetings that proving they really are microfossils and not crystal anomalies of some other sort is impossible.

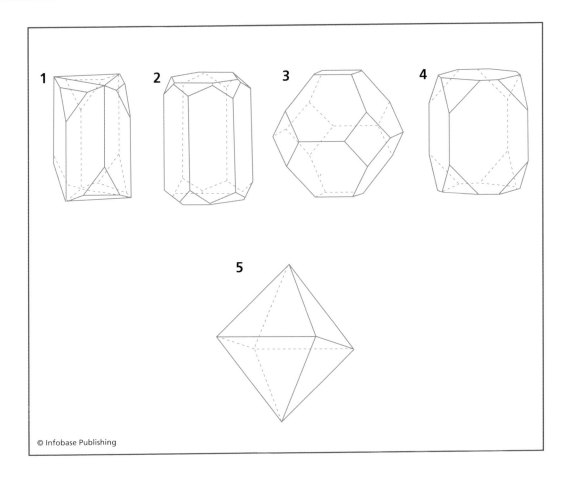

© Infobase Publishing

The carbonate minerals also contain small grains of magnetite that some scientists think were made by bacteria. There are bacteria on Earth that grow tiny magnetite grains that they use as internal compasses. The bacteria use the interaction of these tiny magnetic grains with the Earth's magnetic field to align themselves while feeding in the oceans. Their magnetite grains have a specific crystalline shape not found elsewhere in nature. Scientists studying the magnetite grains from ALH 84001 say they have the same shape as the terrestrial magnetite grains, and so were probably made by bacteria on Mars. The magnetite grains from Mars are much smaller than the terrestrial bacterial grains, with the result that there is no technology available that can make images clear enough to prove that they have exactly the same crystallographic

Subtle differences in the shapes of tiny magnetite crystals found in Martian rocks have been used to argue for the existence of life in Mars's past: Certain shapes of magnetite crystals are only made by bacteria on Earth.

shape as the bacterial grains on Earth, and not a very similar shape that is formed inorganically in nature. There is also no evidence that Mars had a magnetic field long enough for bacteria to benefit from having internal magnetic compasses. Mars does not have a magnetic field now, and though some surface rocks are magnetized, it is unknown how long the field lasted or why it stopped. The carbonate inclusions also contain amino acids that are proposed to have been from life. Some scientists have suggested that the meteorite has been processed too violently while being ejected from Mars and traveling through space to retain any original signature of life on Mars. The rock shows limited and clearly definable evidence of melting from the shock of the impact that ejected it from Mars. In space the rock remained exceptionally cold and unchanged. The heat pulse from Earth atmospheric entry penetrates a meteorite only a few millimeters, so the meteorite interior is fresh. This meteorite has always been cooler than about 100°F (40°C) since leaving Mars. Once on the ice in Antarctica, however, the rock almost inevitably encountered terrestrial life in the form of microbes at the least. Chemical signatures in the carbonate that some scientists believe are evidence of biological activity on Mars are thought by others to be the residue of terrestrial microbes that had migrated through pore spaces deep into the rock.

A higher-level argument also rages in the scientific literature about the formation conditions of the carbonates. Some researchers think that the carbonate compositions and textures require that they formed above 1,200°F (650°C), which is too hot to allow any life-forms known on Earth to live and leave their signatures. Other researchers find that the oxygen isotopic ratios in the carbonates indicate that they formed at less than 570°F (300°C), perfectly compatible with life.

Scientists now focus their search for life on Mars on results from missions to the planet. To decide how to search for life on Mars, they examine life in extreme environments on Earth and attempt to estimate where similar environments might exist on Mars. On Earth single-celled organisms known as extremophiles have adapted to life in otherwise entirely inhospitable places. Some bacteria, called chemolithoautotrophs ("those that live on energy they extract themselves from rock or inorganic chemicals"), live with less than 1 percent oxygen and without sunlight and derive their energy from methane, manganese, iron, or even arsenic. Some eat sulfide minerals and excrete sulfuric acid. Some bacteria live meters or even kilometers deep in bedrock; they may make up half of the Earth's biomass. Bacteria on Earth can live in water with a high-acid pH of 2.3, such as the Rio Tinto River, when most water is around 7; the organism *Ferroplasma* can even live at pH equal to 0. Extremophiles on Earth have been shown to live at temperatures ranging from −4°F (−20°C) to 250°F (121°C), that is, from well below freezing to well above the boiling point of water. On Earth, life has adapted to every environment except higher temperatures still, and extreme dryness. These realizations about the hardiness of life on Earth raises the hopes of scientists looking for evidence of life, or past life, on Mars: Many of the environments Earth organisms have adapted to are matched on Mars.

The search for evidence of life on Mars is fascinating in a multitude of ways. No one knows what life really is: How did early organic molecules become "alive," and what does that mean? If life existed on Mars as well, then perhaps its formation is not as difficult and unlikely as it seems to us now, when Earth is the only planet with life known in the universe. Life on Mars as well as Earth may suggest that life has developed on many planets throughout the universe.

Moons

ars's moons, Phobos and Deimos, are small, irregular satellites entirely unlike Earth's Moon. Their sizes and irregular orbits indicate that they are probably asteroids captured by Mars's gravity field rather than bodies that formed at the same time as the planet. (See the table of information for the moons on page 169.) Asaph Hall, an American astronomer working at the United States Naval Observatory, announced his discovery of Mars's two moons in August of 1877. He named them Phobos and Deimos after the horses that pulled the chariot of Mars in Roman mythology. (Earlier, in 1727 in his book *Gulliver's Travels,* the Irish writer and satirist Jonathan Swift had speculated that Mars had two moons.)

In Greek mythology, Phobos is one of the sons of Ares (Mars) and Aphrodite (Venus). "Phobos" is a Greek word for "fear" and is the root of the word "phobia." Phobos is getting closer to Mars by six feet (1.8 m) per hundred years and will in 50 million years either crash into Mars or break up into a ring. Phobos orbits Mars below the *synchronous orbit radius.* (See the sidebar, "What Are Synchronous Orbits and Synchronous Rotation?" on page 170.) Viewed from Mars, it rises in the west, moves very rapidly across the sky, and sets in the east, usually twice a day. It is so close to the surface that it cannot be seen above the horizon from all points on the surface of Mars. Phobos's significant surface feature is its huge crater Stickney, given the maiden name of Hall's wife, and shown in the image on page 172. Stickney is six miles (9.5 km) in diameter. Like Saturn's moon Mimas with its immense crater Herschel, the impact that caused Stickney must have almost shattered Phobos. Opposite Stickney on Phobos (this is called Stickney's antipodal point) there is a region of ridges and radial grooves. This chaotic terrain is thought to have been caused by the impact that created Stickney: The shock waves moved away from the impact and converged on the other side of the planet. This antipodal point was the focus of a lot of seismic energy coming from all sides, enough to disrupt the crust significantly. Similar antipodal terrains are found on Venus, and possibly on Mars, where the Tharsis volcanic field is antipodal to the Hellas impact basin.

The Soviet spacecraft *Phobos 2* detected a faint but steady outgassing from Phobos. Unfortunately, *Phobos 2* died before

FUNDAMENTAL FACTS ABOUT PHOBOS AND DEIMOS

Parameter	Phobos	Deimos
semimajor axis (from Mars)	5,862 miles (9,378 km)	14,662 miles (23,459 km)
orbital period	0.31891 days	1.26244 days
rotational period	0.31891 days	1.26244 days
orbital inclination	1.08 degrees	1.79 degrees
orbital eccentricity	0.0151	0.0005
major axis radius	8 miles (13 km)	4.7 miles (7.5 km)
median axis radius	7 miles (11 km)	3.8 miles (6.1 km)
minor axis radius	6 miles (9 km)	3.2 miles (5.2 km)
mass	2.3×10^{16} lb (10.6×10^{15} kg)	5.2×10^{15} lb (2.4×10^{15} kg)
mean density	118 lb/ft^3 (1,900 kg/m^3)	109 lb/ft^3 (1,750 kg/m^3)
apparent *visual magnitude*	11.3	12.40

WHAT ARE SYNCHRONOUS ORBITS AND SYNCHRONOUS ROTATION?

Synchronous rotation can easily be confused with *synchronous orbits*. In a synchronous orbit, the moon orbits always above the same point on the planet it is orbiting (this section uses the terms *moon* and *planet,* but the same principles apply to a planet and the Sun). There is only one orbital radius for each planet that produces a synchronous orbit. Synchronous rotation, on the other hand, is created by the period of the moon's rotation on its axis being the same as the period of the moon's orbit around its planet, and produces a situation where the same face of the moon is always toward its planet. *Tidal locking* causes synchronous rotation.

Gravitational attraction between the moon and its planet produces a tidal force on each of them, stretching each very slightly along the axis oriented toward its partner. In the case of spherical bodies, this causes them to become slightly egg-shaped; the extra stretch is called a tidal bulge. If either of the two bodies is rotating relative to the other, this tidal bulge is not stable. The rotation of the body will cause the long axis to move out of alignment with the other object, and the gravitational force will work to reshape the rotating body. Because of the relative rotation between the bodies, the tidal bulges move around the rotating body to stay in alignment with the gravitational force between the bodies. This is why ocean tides on Earth rise and fall with the rising and setting of its moon, and the same effect occurs to some extent on all rotating orbiting bodies.

The rotation of the tidal bulge out of alignment with the body that caused it results in a small but significant force acting to slow the relative rotation of the bodies. Since the bulge requires a small amount of time to shift position, the tidal bulge of the moon is always located slightly away from the nearest point to its planet in the direction of the moon's rotation. This bulge is pulled on by the planet's gravity, resulting in a slight force pulling the surface of the moon in the opposite direction of its rotation. The rotation of the satellite slowly decreases (and its orbital momentum simultaneously increases). This is in the case where the moon's rotational period is faster than its *orbital period* around its planet. If the opposite is true, tidal forces increase its rate of rotation and decrease its orbital momentum.

Almost all moons in the solar system are tidally locked with their primaries, since they orbit closely and tidal force strengthens rapidly with decreasing distance. In addition, Mercury is tidally locked with the Sun in a 3:2 *resonance.* Mercury is the only solar system body in a 3:2 resonance with the Sun. For every two times Mercury revolves around the Sun, it rotates on its own axis three times. More subtly, the planet Venus is tidally locked with the planet Earth, so that whenever the two are at their closest approach to each other in their orbits, Venus always has the same face toward Earth (the tidal forces involved in this lock are extremely small). In general any object that orbits another massive object closely for long periods is likely to be tidally locked to it.

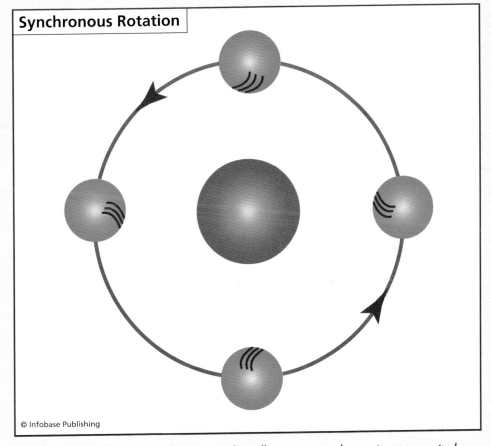

Synchronous Rotation

© Infobase Publishing

In this schematic of synchronous rotation, the yellow moon revolves as it rotates so it always has the same face toward its green planet.

Mars's moon Phobos (NASA/JPL-CalTech/ University of Arizona)

it could determine the composition of the outgassing material; water is a candidate. New images from *Mars Global Surveyor* indicate that Phobos is covered with a layer of fine dust about a meter thick, similar to the regolith on the Earth's Moon.

Deimos's radius is about 3 percent of the Moon's radius and half of Phobos's radius. This tiny moon, named for the Roman god of dread, has a completely black surface covered with small craters. The lack of ejecta streaks from these craters indicates that most of the ejecta probably escaped the tiny body's insignificant gravity field.

Phobos and Deimos may be composed of carbon-rich rock, perhaps the same composition as C-type asteroids. The densities of these moons are so low that they cannot be pure silicate rock and are more likely composed of a mixture of rock and ice. Phobos and Deimos are insignificant in the solar system pantheon of moons, but some scientists think they may someday be useful as "space stations" from which to study Mars or as intermediate stops to and from the Martian surface, especially if the presence of ice is confirmed.

Mars Reconnaissance Orbiter:
A Study in Mission Planning and Science Achievements

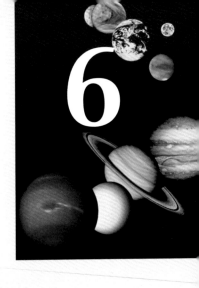

In 1992, an ambitious Mars mission the Mars Observer was launched. This was the United States's first Mars attempt since 1976, when *Viking 2* launched. Between 1976 and 1992, the Soviet Union launched one successful and two unsuccessful missions to Mars's moon Phobos, and the United States wanted to get back into the Mars arena. *Mars Observer,* however, failed: Contact was lost just before the spacecraft was to enter Mars's orbit. A NASA review board concluded that a failure in the propulsion system might have been at fault, but because there were no indications of trouble before contact was lost, this remains a hypothesis.

Though the loss of *Mars Observer* was both expensive and demoralizing, the success of *Mars Pathfinder* and *Mars Global Surveyor* in 1996 helped revive the Mars program at NASA. The data from *Pathfinder* was a scientific bonanza, but it was surface data: None of the global data that *Mars Observer* had been designed for could be obtained with *Pathfinder*. NASA also set out to fly a series of missions to recapture the measurements desired and carry out the original goals of *Mars Observer*. Because that loss had been so costly the new strategy was to avoid putting all the eggs in one basket and instead to fly a series of orbiters. The first was *Mars Global Surveyor,* a hugely successful mission that also launched in 1996 and ran for four times its initial planned lifetime, ceasing communications in November 2006 likely due to battery failure. *Mars Global Surveyor* carried a camera, a laser altimeter, a thermal emission spectrometer, a magnetometer, a radio wave experiment, and a special relay to transmit signals from any future Mars surface missions.

While the next two missions, the *Mars Climate Orbiter* and *Mars Polar Lander,* were on their way to the planet, another orbiter-lander pair was in development. The first of these became the successful *Mars Odyssey* orbiter, which launched in 2001 and continues to provide data after more than 2,700 days in orbit. The second was the *Mars Phoenix* lander, which after many delays launched in August 2007 and landed successfully in May 2008. Its last transmission, right on schedule for its short lifetime, was on November 2, 2008. Meanwhile, *Mars Climate Orbiter* and *Mars Polar Lander* both failed on arrival to Mars. The dual loss was an immense shock and setback to the NASA community. Now the whole Mars program required significant reassessment.

In a way, the *Mars Reconnaissance Orbiter* began with a metric to English conversion error. At the Jet Propulsion Laboratory (JPL) the navigators of the *Mars Polar Lander* were working in metric and at Lockheed Martin, where the spacecraft,was operated, the engineers were working in English units. A meter does not equal a yard, and a kilogram does not equal a pound. As much as everyone involved was entirely aware of these problems, some conversion errors crept in. There were some early indications that the spacecraft was not acting as expected, and some questions had been raised but not soon or loudly enough. It is thought that these conversion errors caused the mission failure. Later some team members felt that if they had been better funded there would have been more oversight and the errors would have been caught. Certainly in general budgets were shaved to the bone, but the larger problem is the failure of the United States to convert to the metric system, as the rest of the world has. In the United States, science is conducted in metric, while much of engineering remains in the English system, creating confusion and additional work in conversions.

NASA began to look again at the whole structure of the Mars program. The U.S. Office of Management and Budget suggested that failures may have occurred because of trying to do too much without enough funding. In the middle of this reassessment came the discovery of the Antarctic Martian meteorite Allen Hill 84001 (ALH84001) and, with it, a serious suggestion that life may have existed on Mars. Though ALH84001 was brought back from Antarctica in 1984, it was not recognized as Martian until 1994, and the first paper discussing the possibility of microfossils was published in 1996.

Though the suspected evidence for life appears not to be so, the excitement around these suggestions was a significant force in planetary science at the time. Within NASA, the questions quickly became three: Is there life on Mars now? Did life exist on Mars in the past? And if not, why not? These questions are closely related to long-standing questions about climate change on Mars. What appear to be dry river valleys are spread across wide regions of the planet. Mars was evidently a wet planet in the past, so why is it dry now? If Mars lost its water and became a desert planet, could Earth follow the same path? And if there was once life on Mars, did the climate change cause its extinction? NASA settled largely on a strategy informally called "follow the water": Where there is water there is the greatest chance of finding life, and so the Mars program was charged with finding out when there was water, where it was, and where it went.

Thus NASA was, in its own words, "rearchitecting" the Mars program both in the immediate aftermath of several devastating mission losses and in the glow of the new excitement over Allen Hills 84001. G. Scott Hubbard from NASA and Charles Elachi, director of the JPL, put together a group to meet and discuss priorities and mission concepts. The people working on the mission plans included Dan Goldin at NASA, Tom Young and others from the Viking mission teams, and Farouz Nadieri from the JPL Mars Office. (The JPL was given and still has a NASA Mars Office, whether or not missions are run from there.) James Garvin, later NASA chief scientist,

(continues on page 178)

WHO DECIDES WHICH SPACE MISSIONS GET LAUNCHED?

When NASA needs to make high-level decisions, for example, about what kinds of space missions should be undertaken, guidance comes largely from three sources: The president of the United States, top NASA administrators, and science advisory groups. Some presidents choose to make space research a priority, as Presidents Johnson and Kennedy did during the Apollo era. Similarly, President George W. Bush encouraged lunar exploration for potential settlement in the future. In 2004, President Bush announced a new Vision for Space Exploration, saying, in part:

> [The Vision for Space Exploration] is a sustainable and affordable long-term human and robotic program to explore space. We will explore space to improve our lives and lift our national spirit. Space exploration is also likely to produce scientific discoveries in fields from biology to physics, and to advance aerospace and a host of other industries. This will help create more highly skilled jobs, inspire students and teachers in math and science, and ensure that we continue to benefit from space technology, which has already brought us important improvements in areas as diverse as hurricane forecasting, satellite communications, and medical devices.

Under the guidance of the presidential directive, a number of advisory groups make recommendations to NASA. Perhaps the most important and influential science advisory groups are those from the National Academies of Sciences and Engineering, in which membership is a high honor given to accomplished scientists by their peers already in the Academy. The National Academies are advisers to government on policy decisions and also serve as oversight committees on governmental affairs. Other important science advisory groups are open to any interested scientists and focus on specific mission goals: the Venus Exploration Advisory Group (VEXAG), the Lunar Exploration Advisory Group (LEXAG), the Mars Exploration Advisory Group (MEPAG), and the Outer Planets Exploration Advisory Group (OPAG).

NASA's Science Mission Directorate, currently under the leadership of Alan Stern, takes these ideas and recommendations and fashions NASA policy from them. NASA's road map for space exploration includes the following science research objectives:

1. How did the Sun's family of planets and minor bodies originate?
2. How did the solar system evolve to its current diverse state?

3. What are the characteristics of the solar system that led to the origin of life?
4. How did life begin and evolve on Earth and has it evolved elsewhere in the solar system?
5. What are the hazards and resources in the solar system environment that will affect the extension of human presence in space?

The road map calls for a series of small (Discovery program, meant to launch one mission every 12 to 24 months), medium (New Frontiers program, meant to launch one mission about every 36 months), and large (flagship) class missions, supported by a balanced program of research and analysis, and creative education and public outreach. These are supplemented by the ongoing Explorers program, which has run a long series of missions set at under $180 million.

In the past, NASA announced plans to explore a certain planet or region of space and solicited independent bids and competitions for spacecraft, operations, and science investigations. These missions were very large in scope, carrying many instruments, involving large groups of people, taking many years to get organized and launched and often costing billions of dollars.

The current movement to smaller missions led by scientists is a significant departure from that earlier management style. Proposals for New Frontiers, Explorers, and Discovery missions are solicited by the publication of NASA's Announcements of Opportunity, in which NASA issues detailed guidelines for mission proposals. Each mission proposal is led by a principal investigator (PI), typically a scientist at a university or research institution.

The principal investigator selects team members from industry, small businesses, government laboratories and universities to develop the scientific objectives and instrument payload. The team brings together the skills and expertise needed to carry out a mission from concept development through data analysis. The PI is responsible for the overall success of the project by assuring that cost, schedule, and performance objectives are met.

Once a PI decides to propose a mission and has a target in mind, he or she needs to find two key partners: one who can build the mission and one who can manage and operate it. These can be the same organization, but more often they are

(continues)

(continued)

separate. The partner to build the mission is usually a defense contractor or other corporation with experience in this area, such as Lockheed-Martin, Ball Aerospace, or AeroAstro. The management and operations partner is often a NASA center, such as the JPL (at the California Institute of Technology), or Goddard-Applied Physics Laboratory (at Johns Hopkins University). The PI also invites a number of scientists to be on the science team for the mission, and together they decide which measurements they want to make and then work with the engineers to determine how these measurements might be made.

Even the proposal phase of a mission is immensely time-consuming and expensive. It involves multiple trips to meet with the team, many "telecoms" (telephone conferences), and a considerable amount of data analysis, budgeting, and even building instruments, testing, and proving feasibility. For example, to prove to the scientists who assess the multiple proposals fighting for selection that an instrument will work on Venus, it may be necessary to find a lab somewhere in the world where the instrument can be placed in a chamber to make measurements under Venus surface conditions, at 855°F (457°C) and about 100 times Earth's atmospheric pressure. The resulting proposal is usually hundreds of pages long. It is likely to be rejected, as there are more proposals by far than available mission budgets, and no scientists are paid to write proposals.

The Discovery program has existed for some years now and has launched 10 missions to date (Near Earth Asteroid Rendezvous in 1996; Mars Pathfinder in 1996; Lunar Prospector in 1998; Stardust in 1999; Genesis in 2001; Comet Nucleus Tour, or CONTOUR, in 2001; MESSENGER in 2004; Deep Impact in 2005; Dawn in 2007 and Kepler in 2009).

The Space Studies Board, a committee within the National Academies, first issued reports recommending that the New Frontiers program be established and then which planetary targets should be considered when choosing missions for the New Frontiers program. In their first recommendation, the Space Studies Board listed Venus, Jupiter, the south polar region of the Moon, and outer solar system objects as desired targets

(continued from page 175)

was then chief NASA scientist for Mars exploration, and he brought a lot of energy and fresh thinking to this process.

This core group solicited ideas from a wide range of independent scientists and engineers in an attempt to create a connected, consistent series of missions that would best serve for exploration. New Frontiers's first mission launched was New Horizons, the mission for outer solar system research that left Earth in 2006. In 2011 the mission Juno is expected to launch, with the aim to conduct in-depth observations and measurements of Jupiter.

The New Frontiers program seeks to contain total mission cost and development time and improve performance through the use of validated new technologies, efficient management, and control of design, development, and operations costs, while maintaining a strong commitment to flight safety. The cost for the entire mission must be less than $700 million. In their most recent recommendation, in spring 2008, the targets have been broadened so greatly that any high-quality science-driven mission may be considered; the call for proposal has been made, and the process of selection for the next New Frontiers mission will begin in 2011.

There is a lot of emphasis right now at NASA on who gets to be a PI; no longer can any scientist interested in a mission manage to cobble together a team and send in a proposal. In the Discovery and New Frontiers programs, the PIs have a huge role, far larger than existed in older missions when there were no PIs in the same way. The PI–led missions have tended to have cost overruns, so NASA is more concerned about having someone experienced. Prospective PIs are now required to go to a PI summer school run by NASA and to qualify through an online survey.

Most important, however, the space sciences community is small, and team members and PIs are likely to be known by anyone else in the community. Publishing scientific papers, attending conferences, and participating in advisory councils all create name recognition if not personal friendships. For anyone interested in space science, showing up and getting to know the community is the most important first step, along with conducting interesting science and publishing it. This process can begin in college or even in high school—interested high school students can get summer jobs in science labs at universities, and in small numbers each year they do attend scientific conferences, hitting the ground running.

the science goals. This process was highly successful and led in the end to an ongoing series of meetings and papers from a group now called MEPAG, the Mars Exploration Program Analysis Group. These meetings are chartered by NASA but open to all who are interested. Other planets and bodies now have their own independent mission steering committees: For example, Venus has VEXAG, the Venus Exploration Analysis Group, and the Moon has LEAG, the Lunar Exploration Analysis Group.

For orbital data, the Mars team wanted to get higher resolution imaging data, in two flavors: much higher resolution in smaller areas or more widespread coverage at a lesser resolution. This tradeoff in how to spend limited data bits became an ongoing argument in the meetings. The *Mars Global Observer* data was at best one yard (1 m) per pixel, but often as much as four yards (4 m) per pixel, and it came in narrow strips that made regional interpretations difficult. These narrow strips became critical in planning what then became the *Mars Reconnaissance Orbiter* concept (then called the *Mars Science Orbiter,* which is confusing because there is a future mission with that name).

During the spring and early summer of 2000, a number of mission concepts were advanced. NASA said we needed to get back to the surface, but we also needed in-depth reconnaissance from orbit. Two concepts then took the lead in head-to-head competition for the 2003 launch opportunity (the ability to fly from the Earth to Mars requires that the planets be in certain points in their orbits, and in general a Mars launch opportunity arises every two years). The first was the Mars Reconnaissance Orbiter mission, and the other was a novel idea about a lander being placed in air bags. This became the Mars Exploration Rover missions, led by Steve Squyres from Cornell University. The Rover concept was to take the lander payload from a planned 2001 mission that was cancelled; having instruments that were already either built or approved is a huge advantage in mission planning. The *Mars Reconnaissance Orbiter* (still known then as *Mars Science Orbiter*) was led by Rich Zurek from NASA. Zurek had the idea to get the instruments that had been lost back in place.

Clearly the Rover missions were not going to the poles. The Mars Polar Lander had been an unusual, special project, made possible because the declinations of Mars and Earth lined up at that period in a way that made polar landing possible, and Mars's climate at that time had a maximum atmospheric pressure at a time when the pole was in sunlight, an unusual coincidence of events making it possible to land on the south pole. The new Phoenix lander mission, for example, had to land near the north pole.

In response to knowing the landing constraints on the Rover missions and the loss of such hoped-for data from the 1998 mission, Zurek and his team put together a number of ideas for instruments, including a high-resolution imaging system with sub-meter spatial resolution.

The two competing missions came head-to-head at NASA headquarters, presenting to a committee chaired by Scott Hubbard to review the scientific and technical aspects of the proposals. At that time the Rover had two advantages. One was that the Rover had an existing payload, with hardware already tested, approved, and built. The second was that it was thought that the 2005 launch to Mars opportunity would be more difficult for direct entry for the rover because of the relative positions of the planets and Mars's eccentricity. Together the planets' positions at the time of landing would create higher entry velocities, requiring stronger and larger aeroshields to protect the craft as it entered Mars's atmosphere. As tenuous as it is, the atmosphere of Mars can significantly heat and damage entering spacecraft. Because the budget and launch vehicle determine the highest possible mass for the craft, any additional mass needed to build stronger and larger aeroshields takes mass away from the mission payload, its science instruments. Launching Rover in 2005 meant the mission would have less science.

The discussion on Mars Reconnaissance Orbiter centered on the science instrumentation decisions: Does the space community really agree on what is wanted? Some scientists still argued that the highest possible resolution was needed, while others maintained that images in narrow strips are not useful. This controversy raised the slight doubt that the mission was not as ready to go forward. In addition, because not all of the orbiter's instruments were ready to go, NASA would have to release an Announcement of Opportunity (an AO), asking for science and engineering teams to propose instruments to build. This is a time-consuming process, from writing and releasing the AO to waiting for the proposing teams to form, create ideas, plans, and budgets, and submit formal proposals, to choosing the winners, and finally actually building and testing the new instrument. The committee decided: The Rover mission would take the earlier launch opportunity and Mars Reconnaissance Orbiter would plan for the 2005 launch.

Disappointing as it was not to be chosen to fly first, the extra time was valuable for the Mars Reconnaissance Orbiter team. While designing the AO the team decided not, in fact, to legislate the resolution issue: Why choose? One imager will be called a "context imager" and give six yards (6 m) per pixel in a 19 mile- (30 km-) wide swath of the surface, and then a second very high-resolution camera will take images over smaller areas. The other instruments quickly came into line, with the goals of recovering the objectives of the *Mars Climate Orbiter* while continuing the ongoing work of Surveyor.

The AO was released. The instrumentation AOs are released onto NASA's Web site and are open to any interested groups. This is a request for proposals to build instruments and lists minimum capability requirements for the instruments (for example, a range of wavelength of energy to measure, or a per pixel resolution requirement), with an overall allocation of energy and mass for the spacecraft and a breakdown by instrument. Since the spacecraft has to provide all the power used by the instruments to make their measurements and send their data back to Earth, energy guidelines are at least as significant as those for mass and volume. These instruments tend to be very small and very efficient. This is one of the ways that NASA ends up producing new technologies for people on Earth to use as well.

Typically two or three proposals are submitted for each instrument. The proposals come most often from scientists at universities, called in this circumstance Principal Investigators for the instruments, paired with industry partners, such as Ball Aerospace or Ames Aerospace, or with one of the strategic NASA centers (Ames Research Center,

The Mars rover Opportunity *can be seen in this HiRISE Mars Reconnaissance Orbiter image of Victoria Crater in Meridiani Planum. The crater is about a half mile (0.8 km) across and has an unusual scalloped rim. (NASA)*

Dryden Flight Research Center, Glenn Research Center, Goddard Space Flight Center, Jet Propulsion Laboratory, Johnson Space Center, Kennedy Space Center, Langley Research Center, Marshall Space Flight Center, Stennis Space Center, and Wallops Flight Facility) with the special engineering expertise required. The AO also brought in excellent proposals. Various suites of payload instruments were considered, and in the end, the climate side of measurements went to the Infrared Climate Sounder and the MARCI wide field camera that does daily weather photos. Now it seems obvious that a very high-resolution camera like Hi-Rise camera is necessary to look for landing spots, understand the distribution of rock sizes on the surface, to look for and check on on current surface missions, and make images to test hypotheses of surface changes, but it was not obvious at the time. Some scientists said that to justify putting a very expensive camera on the mission it had to promise a big leap in science, only possible with resolution at least 10 times better than what had flown before. This camera was recommended by the science definition team, along with the CRISM spectrometer, so there was a big drive toward near IR spectroscopy and what it could do for compositional information.

The mission thus had an excellent list of instruments, but the launch vehicle and other mass and volume constraints showed they could add one more. They had the capacity to add an instrument but reached the limit of the budget. The Italian Space Agency in coordination with NASA decided to create a radar experiment to fly on *Reconnaissance Orbiter*. This shallow surface radar instrument, SHARAD, was sold to the mission team on the basis of looking for subsurface water ("follow the water"). This instrument can penetrate the soil surface of Mars with less depth but more resolution, offering the possibility of a far more detailed understanding of the existence of frozen water in Mars's surface soils. SHARAD has an Italian Principal Investigator with an American Deputy Principal Investigator. With that addition, the payload of the orbiter was complete.

Not only was the camera higher resolution but so were the atmospheric sounder and the imaging spectrometer. The craft therefore needed a higher data transfer rate, and that created specific requirements for a 100 watt traveling waveguide amplifier and a 9.8 foot (3 m) mirror. Adding to the complications of these planning steps, the Lockheed Martin spacecraft was selected before the team had determined the specifics of the payload. Some of the instrumentation proposals had surprises, like the innovation of a color stripe down the center of the high-resolution camera (Hi-Rise) images. In the end the mission was complete with six instruments:

* HiRISE, the High Resolution Imaging Science Experiment, a camera that takes very high resolution surface images;
* CTX, the Context Camera, which provides wide-area views;
* MARCI, the Mars Color Imager, takes weather images to monitor clouds and dust storms;
* CRISM, the Compact Reconnaissance Imaging Spectrometer for Mars, which looks at specific wavelengths of energy from the Mars surface to identify mineral types;

* MCS, the Mars Climate sounder, detects vertical variations in temperature, dust, and water vapor in the Mars atmosphere; and
* SHARAD, the Shallow Radar experiment, which probes the Martian surface seeking water at depths greater than 3.3 feet (1 m).

These six instruments are complemented by a gravity science team that uses the mass of the spacecraft itself to make measurements as it responds to the planetary gravity field and a science group that uses the response of the spacecraft to aerobraking, dipping into the highest Martian atmosphere, to investigate characteristics of that atmosphere.

Similarly NASA missions usually hold a competition for the industry partner for the mission as a whole and in parallel with the instrumental AO a request for proposals for the spacecraft went out to industry. For the launch and craft itself, NASA has really only a few vendors that can compete. Like the Rover project, the Reconnaissance Orbiter was being managed by the JPL. The JPL can make vehicles (this is called an "in-house build"), but they were already dedicated to the Rover project and had no more capacity. The regular industry partners are Lockheed Martin, Orbital Sciences Corporation, Ames Aerospace, Ball Aerospace, and to a lesser extent, Boeing Corporation. The knowledge of how to build and launch space missions is a critical one and can be lost if the people with the knowledge are not kept employed and their knowledge used. NASA likes to have several private corporations with this knowledge, in addition to the experts at the 11 NASA centers, but the winning proposal is always chosen based on merit. Lockheed Martin won the *Reconnaissance Orbiter* contract with an innovative design that took into consideration the lessons learned from the failed missions. In general, Lockheed Martin has an advantage in these competitions because they have worked on so many missions in the past.

Once the instruments and the craft were selected, the team began to talk more about the orbits the mission would take. *Mars Reconnaissance Orbiter* was going to a lower altitude orbit to increase resolution (the closer the craft is to the surface, the higher resolution images can be without more sophisticated engineering). The lower orbit introduced certain problems, such as planetary protection—the spacecraft has to be really clean in case it breaks up and falls to the planet. NASA works very hard to prevent Earth microorganisms from being introduced onto planets visited by space missions. Many microbes, unfortunately, can survive the long trip through the freezing vacuum of space, and so exceptional precautions are taken on Earth to keep the craft clean before launch.

A special triumph was made in the orbit selection because it gives an interesting repeat cycle, which when combined with the spacecraft's ability to roll to right and left, allows viewing of any target of Mars within a 17-day period. This rapid scanning of a wide portion of the planet's orbit makes planning so much easier: In the regular mission meetings, if two science team's requests are conflicting, they both can be fulfilled within 17 days. The mission's orbit is circular, which keeps it at a relatively constant altitude above the planetary surface. The circular orbit makes creating mosaics of the surface much easier. If the altitude from which the image is taken constantly varies, some sophisticated math has to be used to adjust the images to one common altitude before they can be stitched together into a mosaic. Additionally, the repeat approaches to the same regions make production of stereo images easy—these are the images that when viewed with special glasses make the surface look three-dimensional. They require images of the same place from two different angles.

The mission thus has engineers to run the craft at the JPL and at Lockheed Martin, and they have teams of scientists. Some scientists are at the mission level: Rich Zurek, who is the lead project scientist, and Suzanne Smrekar, deputy project scientist, and Jim Graf, project manager, all work at JPL along with a small additional team of engineers and scientists. In general, though, the scientists who guide the instruments and make the first interpretations of the data come along with the instruments themselves; they were involved in the proposal and have stayed on through the long years to launch and to arrival at Mars. Their instruments on *Mars Reconnaissance Orbiter* will form just a fraction of what they do in their work lives, which are generally at universities or at NASA centers. All these scientists will receive minor funding for the science on the mission and most of their support will be for other scientific research and teaching.

As the mission neared Mars, however, the team put out an additional request for proposals, in this case not for more hardware but for more people. This time scientists could write proposals for research they would like to do related to the goals of the Reconnaissance Orbiter mission and tied to data that the instruments are likely to obtain. These new scientists are called participating scientists, and each is assigned to a specific instrument so they have a home team. The NASA system is remarkably open and accessible; as an example, the following is the introductory text to the original AO for the Mars Reconnaissance Orbiter participating scientists:

> This NASA Research Announcement (NRA), entitled *Research Opportunities in Space and Earth Sciences (ROSES)—2006,* solicits basic and applied research in support of the Science Mission Directorate (SMD), National Aeronautics and Space Administration (NASA). This NRA covers all aspects of basic and applied supporting research and technology in space and Earth sciences, including, but not limited to: theory, modeling, and analysis of SMD science data; aircraft, stratospheric balloon, and suborbital rocket investigations; development of experiment techniques suitable for future SMD space missions; development of concepts for future SMD space missions; development of advanced technologies relevant to SMD missions; development of techniques for and the lab analysis of both extraterrestrial samples returned by spacecraft as well as terrestrial samples that support or otherwise help verify observations from SMD Earth system science missions; determination of atomic and composition parameters needed to analyze space data as well as returned samples from the Earth or space; Earth surface observations and field campaigns that

support SMD science missions; development of integrated Earth system models; development of systems for applying Earth science research data to societal needs; and development of applied information systems applicable to SMD objectives and data. . . . Awards will be made as grants, cooperative agreements, contracts, and inter- or intra-agency transfers depending on the nature of the proposing organization and/or program requirements. The typical period of performance for an award is three years, although a few programs may specify shorter or longer (maximum of five years) periods. Organizations of every type, domestic and foreign, Government and private, for profit and not-for-profit, may submit proposals without restriction on number or teaming arrangements. Cost sharing is encouraged but not required. Note that it is NASA policy that all programs involving non-U.S. participants will be conducted on the basis of no exchange of funds.

Though most scientists are concerned with the data and interpretations of their own instruments, the scientists at the JPL also provide mission management. Dr. Zurek and Dr. Smrekar spend a significant amount of time coordinating operations. There are five different instruments teams with competing desires for where the spacecraft should point and how power should be allocated. *Mars Reconnaissance Orbiter* is unique in that the team can roll the spacecraft to point the context camera, CRISM, and the high-resolution camera, Hi-RISE, at specific targets . . . Thus the teams still need to meet weekly to discuss changes in software, data exchange, pointing, and how to adjust the craft to make new kinds of measurements. Every three months the team holds a director's review, where Dr. Zurek or Dr. Smrekar meet with each instrument

---

(opposite page) The spectra of surface minerals that were detected by the CRISM instrument on the Mars Reconnaissance Orbiter. The shapes of these spectra match the minerals labeled here. All the minerals in the top graph contain both sulfur and water and could have been formed in oceans or by flowing groundwater. All the minerals in the bottom graph are formed by reacting iron-bearing minerals with water. (JPL/APL/NASA)

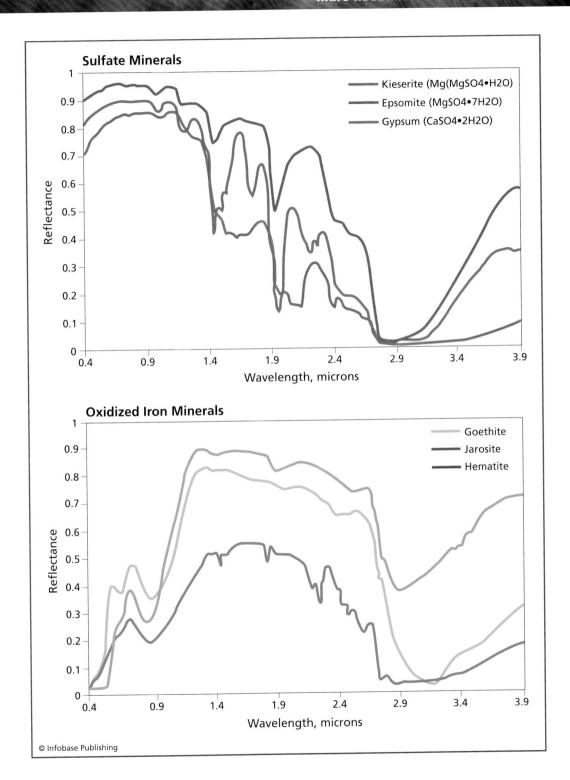

© Infobase Publishing

team to hear about progress and problems. The JPL team also keeps up the Web site and handles the budget, strategizing about how to handle the yearly 5 percent minimum budget cuts. And the overall work concerns keeping all the scientists collaborating and making joint observations in line with the mission's science objectives. Much of the mission is made of management, the same kind required by industry or any other joint human endeavor.

Mars Reconnaissance Orbiter entered orbit around Mars in March 2006 and spent the next six months aerobraking (slowing by using the friction of the spacecraft on the planetary atmosphere) into a circular orbit. The orbiter joined three other orbiting missions: the Mars Global Surveyor, Mars Odyssey, and the European Space Agency's Mars Express. The craft reached its final orbit in November 2006 and has been sending back data since then. Of the many science results that have come from this data interpretation, two are particularly significant: First, the imaging spectrometer has indicated that there is a diversity of aqueous mineral environments on Mars, that is, places where surface water altered the rocks and soils to contain water in their mineral structures. The soil water content shows that in different places water behaved in different ways, and that there were periods of wetness and dryness that alternated. The *Mars Reconnaissance Orbiter* data showing water variations across the surface in both space and time are richly complementary to small-scale observations from the *Mars Exploration Rover*'s results for local water interaction.

The higher-resolution images of the surface made possible by the *Mars Reconnaissance Orbiter*'s instruments show a far more complex planet than has been seen in the past. An example of the water interactions the mission has detected are polygons in the ground. On Earth, similar polygons are created by water expansion during freezing events in soil or rocks or by removal of groundwater through freezing and sublimation (conversion from ice to gas). On Mars, some polygons are fractures in rock, where groundwater has altered the minerals in the rock and wedged the rock into fractures as it froze, and others are polygon shapes created by movement of loose soil and rocks by ice.

The second great result from this mission concerns the patterns and variations in ice on the planetary surface. The orbiter imaged stacks of layers on the poles, but they are not uniform, they are periodic: Their patterns relate to how sunlight strikes the poles. The angle of sunlight is controlled by the tilt of the planetary pole, the eccentricity of Mars's orbit, and other details of its dynamics. Here on Mars, where there is no large moon to stabilize its orbit, rapid and violent climate change was created by orbital changes. The layers in the caps were shown by this mission to be cap-wide and not local, and they are all thought to have been formed in the last 5 million years, an eye-blink in geological time.

Further, the orbiter has detected ice deposits buried by debris located near the planet's equator. These equatorial glaciers could not have formed in the current Mars climate, where ice is not stable at the equator. They had to have formed in the recent past at a time when Mars was far colder, and they are only preserved by layers of insulating rocks and soil. They prove that Mars's inclination has changed radically and recently; some models show that Mars may have been tipped almost on its side in the recent past.

The overall story from this mission is one of planet-scale climate change. In general, the oldest crustal rocks contain phyllosilicates, minerals that incorporate water into their crystal structures. More recent rocks contain sulfates but no water, indicating a move from a watery to an acidic environment. The *Mars Reconnaissance Orbiter* instruments see far more complex changes. There are places on the surface where water persisted far longer than average and places where water came and went in cyclical

This false-color image from the HiRISE camera on the Mars Reconnaissance Orbiter shows the north polar–layered deposits at top and darker materials at bottom, exposed in a scarp at the head of Chasma Boreale, a large canyon eroded into the layered deposits. (NASA/JPL-CalTech/University of Arizona)

patterns and places where the acidity of the soil varied widely. In a few places, the mission has even detected opals, which can only be formed in the presence of water.

The primary mission phase of *Mars Reconnaissance Orbiter* was two Martian years, which ended in the fall of 2008. The mission is now in its first extended phase. CRISM's coolers are beginning to have errors and will not last forever, but for the moment the mission is healthy. Team scientists are now publishing the new science results: More than 50 peer-reviewed papers in scientific journals came out in the last year. With an orbiter, time is needed to build up data from all over the planet to make inferences about change on a planetary scale.

The mission will continue with reduced budgets as long as possible, probably limited by the ability to run the mission on less and less money and not by the hardware itself. With the great success of the Exploration Rovers and other missions, this has become a pattern: As missions continue well past their design lifetimes because of excellent engineering and careful management, budgets run out. New money and even mission management space is needed for newer missions, and old missions must make way. As Dr. Smrekar, the deputy mission scientist says, we won't know the threshold of how much money is too little to operate safely until we hit that threshold. In the meantime, the *Mars Reconnaissance Orbiter* will continue to send back data on global climate change from another planet.

Missions

The conspicuous successes of the Mars Exploration Rover A *(Spirit)* and Rover B *(Opportunity)* revived the Mars program in NASA and paved the way for the *Mars Reconnaissance Orbiter, Phoenix,* and the planned *Mars Science Laboratory* (recently renamed *Curiosity*). The ambitious Mars Science Laboratory mission is producing new technology that will enable later future missions as well. Foremost among the new technology is the Sky Crane. The Sky Crane is a new landing method, required because the *Mars Science Laboratory* exceeds the mass limits of the balloon bouncing landing system that the Mars Rovers used. The *Mars Science Laboratory* will come in with a parachute to slow entry, and then rockets will slow the landing craft, eventually holding it stationary as the rover itself is lowered to the Martian surface on a platform by a crane.

The success of this new landing technology will allow NASA to collaborate more fully with other space agencies and potentially land more than one craft on the planet at a time. This capability is a central consideration in the planning process of the National Academies Decadal Survey, in

(continues on page 196)

PAST AND POSSIBLE FUTURE MISSIONS TO MARS

Launch Date	Mission	Country	Comments
October 10, 1960	Marsnik 1 (Mars 1960A)	USSR	Attempted Mars flyby (launch failure)
October 14, 1960	Marsnik 2 (Mars 1960B)	USSR	Attempted Mars flyby (launch failure)
October 24, 1962	Sputnik 22	USSR	Attempted Mars flyby
November 1, 1962	Mars 1	USSR	Mars flyby (contact lost)
November 4, 1962	Sputnik 24	USSR	Attempted Mars Lander
November 5, 1964	Mariner 3	USA	Attempted Mars flyby (failed)
November 28, 1964	Mariner 4	USA	Mars flyby
November 30, 1964	Zond 2	USSR	Mars flyby (contact lost)
July 18, 1965	Zond 3	USSR	Lunar flyby, Mars test vehicle
February 25, 1969	Mariner 6	USA	Mars flyby
March 27, 1969	Mariner 7	USA	Mars flyby
March 27, 1969	Mars 1969A	USSR	Attempted Mars orbiter (launch failure)
April 2, 1969	Mars 1969B	USSR	Attempted Mars orbiter (launch failure)
May 8, 1971	Mariner 8	USA	Attempted Mars flyby (launch failure)
May 10, 1971	Kosmos 419	USSR	Attempted Mars orbiter/lander
May 19, 1971	Mars 2	USSR	Mars orbiter/attempted lander (failed)
May 28, 1971	Mars 3	USSR	Mars orbiter/lander (failed)
May 30, 1971	Mariner 9	USA	Mars orbiter; first artificial satellite of Mars
July 21, 1973	Mars 4	USSR	Mars flyby (attempted Mars orbiter)

Launch Date	Mission	Country	Comments
July 25, 1973	Mars 5	USSR	Mars orbiter (failed)
August 5, 1973	Mars 6	USSR	Mars lander (contact lost)
August 9, 1973	Mars 7	USSR	Mars flyby (attempted Mars lander)
August 20, 1975	Viking 1	USA	Mars orbiter and lander; first successful landers on Mars
September 9, 1975	Viking 2	USA	Mars orbiter and lander
July 7, 1988	Phobos 1	USSR	Attempted Mars orbiter/Phobos landers
July 12, 1988	Phobos 2	USSR	Mars orbiter/attempted Phobos landers
September 25, 1992	Mars Observer	USA	Attempted Mars orbiter (contact lost)
November 7, 1996	Mars Global Surveyor	USA	Mars orbiter
November 16, 1996	Mars 96	USSR	Attempted Mars orbiter/landers
Dec. 4, 1996	Mars Pathfinder	USA	Mars lander and rover
July 3, 1998	Nozomi (Planet-B)	Japan	Mars orbiter
Dec. 11, 1998	Mars Climate Orbiter	USA	Attempted Mars orbiter
Jan. 3, 1999	Mars Polar Lander	USA	Attempted Mars lander
Jan. 3, 1999	Deep Space 2 (DS2)	USA	Attempted Mars penetrators
April 7, 2001	2001 Mars Odyssey	USA	Mars orbiter
June 2, 2003	Mars Express	USA, ESA, ISA	Mars orbiter and lander

(continues)

PAST AND POSSIBLE FUTURE MISSIONS TO MARS
(continued)

Launch Date	Mission	Country	Comments
June 10, 2003	*Spirit (Mars Exploration Rover A)*	USA	Mars rover
July 7, 2003	*Opportunity (Mars Exploration Rover B)*	USA	Mars rover
August 10, 2005	*Mars Reconnaissance Orbiter*	USA	Mars orbiter
August 4, 2007	*Phoenix*	USA	Mars scout lander
Fall 2011	*Mars Science Laboratory (recently renamed Curiosity)*	USA	Mars science laboratory rover
2013	*MAVEN*	USA	Mars scout mission

(continued from page 193)

which panels of scientists and engineers meet to create proposed NASA mission plans for the next 10 years. For Mars, international collaboration and use of the Sky Crane are expected to initiate a new period in exploration, likely to include steps toward sample return and a planetwide geophysical network.

Mariner 4 1965
American

Mariner 4 was a flyby mission to Mars. It sent back images showing that Mars is no longer an active planet and is heavily cratered.

Mariner 6, 7 1969
American

Flybys over south pole and equator, analyzing atmosphere and taking surface photos. By chance, this mission missed the huge volcanic centers of Tharsis and the giant canyon Valles Marineris.

Mariner 9 1971
American

Mariner 9 was the first artificial satellite of Mars. This mission recorded the first giant Martian dust storm seen by man and supplied close-up photos of Mars and its moons.

Viking 1, 2 1976
American

Viking 1 was launched on August 8, 1975, and landed in Chryse Planitia ("Plain of Gold") on July 20, 1976. These orbiters made global maps and had the first successful landers on Mars, which took photographs of the surface and did experiments on the soil to test for the possibility of life. *Viking*'s photos were the first sent back from the surface of Mars. The Viking missions cost the current equivalent of about $3 billion, an immense investment.

Phobos 2 1989
Soviet

This orbiter took photos of Phobos. Two other Phobos probes failed in the summer of 1988.

Mars Observer 1993
American

Contact was lost with the spacecraft just as it was about to enter orbit around Mars. Because there were no indications of problems aboard the spacecraft prior to loss of contact, no one knows what happened with certainty. A NASA review board spent several months investigating the loss and concluded that the most plausible theory was that a critical failure in the propulsion system disabled *Mars Observer*. Steps were taken to correct the surmised problem in *Mars Global Surveyor*, which carried the same instruments. *Mars Observer* likely flew past Mars and is now in an orbit around the Sun.

Mars Pathfinder 1996
American

Mars Pathfinder was a lander, carrying the *Sojourner* rover. This mission succeeded in making the first "bouncing" landing, and the rover made in-situ chemical analyses of rocks on the surface of the planet. The mission lasted three times longer than expected, from July to September 1997, and produced the first chemical analyses of surface rocks on Mars.

Mars Global Surveyor 1997
American

In 2004, *Mars Global Surveyor* completed its 25,000th orbit of Mars. This successful mission has mapped mineralogy on the surface, monitored weather, and made the first measurements of Mars's weak magnetic field. *Surveyor,* a large craft, measures

Engineers test huge, multilobed air bags, which were designed to envelop and protect the Mars Pathfinder spacecraft before it hit the surface of Mars. The bags measure 17 feet (5 m) in diameter. (NASA/ GRIN)

"Yogi" is a yard-size (meter-size) rock about 15 feet (5 m) northwest of the Mars Pathfinder lander and was the second rock visited by the Sojourner rover's alpha proton X-ray spectrometer instrument. (NASA/JPL)

about 33 feet (10 m) from tip to tip. *Surveyor*'s large high-gain antenna allows fast data transfer to Earth, as rapid as 85,333 bits per second (compare this modern rate to *Mariner 4*'s 8.3 bits per second). In total, the mission has sent back at least 80 gigabytes of data to Earth. The mission's instruments are listed here:

* Mars Orbital Camera (MOC): The camera produces a daily wide-angle weather image of Mars. The narrow-angle lens has a resolution as small as five feet (1.5 m) across.
* Mars Orbiter Laser Altimeter (MOLA): The altimeter bounces a laser off the surface of the planet and has produced a high-resolution, high-accuracy topographic map of the entire planet. MOLA topography of Mars has a higher resolution than topographic data for many areas on Earth.
* Thermal Emission Spectrometer (TES): By measuring electromagnetic emissions from the surface of the planet in the infrared range, scientists can estimate the compositions of surface materials.

✳ Magnetometer/Electron Reflectometer: The magnetometer has made the first detailed measurements of Mars's enigmatic magnetic field.

✳ Mars relay: The relay is an antenna designed to transmit to Earth data received from any future missions that land on the Martian surface.

✳ Radio science: By sending radio signals to Earth from Mars orbit, the shape of the planet and the structure of the atmosphere have been measured.

Mars Global Surveyor operated in orbit for nine years and 52 days, second only to *Odyssey* for the longest mission. Data

Jet Propulsion Laboratory workers in the Payload Hazardous Servicing Facility prepare the Mars Global Surveyor spacecraft for transfer to the launchpad by placing it in a protective canister. (NASA/JPL/ GRIN)

from this mission showed the bright new deposits in gullies in the sides of craters that spawned a whole era of scientific conjecture about bursts of Martian groundwater and motivated a series of measurements by *Mars Reconnaissance Orbiter*. It measured regions of the surface rich in hematite, an iron-bearing mineral that forms in wet environments, and motivated the Mars Rover teams to choose a hematite-rich site for the rover *Opportunity*. Its MOLA instrument produced data that made a topographic map more detailed and precise than is available for Earth. It located 20 impact craters on the Martian surface that are less than seven years old. The *Mars Global Surveyor*'s last communication with Earth was on November 2, 2006.

Mars Polar Lander 1998
American

This mission was lost on arrival to Mars in September 1999.

Nozomi 1998
Japan

On January 3, 1998, Japan launched the *Nozomi* probe to Mars, the first planetary mission by a country other than the United States or the Soviet Union/Russia. Using a combination of lunar gravity, Earth gravity, and rocket burns, *Nozomi* was scheduled to arrive at Mars in December 2003. After five years of travel through the solar system the mission goal of Mars was abandoned when the probe was found to be too far off course to successfully enter a stable Mars orbit without risking a crash onto the surface. (Being an orbiter, the probe had not been built to strict non-contamination standards and a crash would have risked introducing Earth microbes to Mars.) The probe may still be used to measure solar activity as it orbits the Sun.

Mars Climate Orbiter 1999
American

Lost on arrival to Mars in September 1999 due to confusion between NASA and a contractor over the use of feet or meters as units of measurement. The loss of the *Mars Climate Orbiter* and the *Mars Polar Orbiter,* two immensely expensive missions, almost ended Mars exploration by the United States.

Mars Odyssey 2001
American

The Mars Odyssey mission was originally planned as a dual mission consisting of a lander called the *Mars Surveyor 2001 Lander* and a paired orbiter, the *Mars Surveyor 2001 Orbiter.* During the major reorganization of the Mars program that occurred after the losses of the *Polar Lander* and *Climate Orbiter,* the lander portion of this dual mission was cancelled and the orbiter was renamed the *Mars Odyssey.*

Mars Odyssey is a remote-sensing orbiter designed to map chemical elements and minerals, look for water in the shallow subsurface, and analyze the radiation environment. The mission is creating a complete day, night, and visible light map of the whole planet. *Mars Odyssey* carries several kinds of spectrometers:

* Thermal Emission Imaging System (THEMIS): This camera images Mars in the visible and infrared parts of the electromagnetic spectrum to determine the composition and distribution of minerals on the surface of Mars.
* Gamma Ray Spectrometer (GRS): This camera uses the gamma-ray part of the electromagnetic spectrum to look for the presence of 20 elements, including carbon, silicon, iron, and magnesium. Its neutron detectors look for water and ice in the soil.
* Martian Radiation Experiment (MARIE): This instrument measures the radiation environment of Mars using an energetic particle spectrometer. Ironically, the MARIE instrument was disabled in 2003 by an intense radiation burst from a solar coronal mass ejection. The same radiation burst caused auroras on Earth as far south as Florida.

Odyssey's primary science mission began in February 2002 and continued through August 2004. The craft acted as a communications relay for the Mars rovers and helped plan landing sites for future missions. The mission successfully characterized the radiation environment of Mars, critical to paving the way for astronauts to visit Mars. The MARIE experiment found that Mars has two to three times the radiation found near Earth. The mission also detected very high levels of the element potassium on the planet's surface, which along with the abundance of sulfur indicates that water transported material around the surface. Much of the early evidence for water activity in Mars's history was measured by *Odyssey,* including the critical finding that beyond 60 degrees latitude on both poles the Martian surface appears to be well over 50 percent water ice by volume. This finding was great support for the landing site chosen for the Phoenix mission.

Odyssey continues to function as it orbits Mars. Engineers rebooted the system, a risky procedure when in space, in 2003 and in spring 2009, with excellent results including the restoration of a backup drive. The mission is now the longest-serving craft around Mars.

Mars Express 2003
American, European Space Agency, Italian Space Agency

Mars Express orbits over the Martian poles looking for water and at surface conditions. Its lander, the *Beagle* (named for the ship that carried famed naturalist Charles Darwin on his Pacific Ocean explorations), was lost when it hit the surface. No transmissions were received from it, and its exact fate is unknown. *Mars Express* carries a number of instruments designed to detect compositions and other attributes of the Martian surface:

* The High-Resolution Stereo Camera (HRSC): The unusually high-resolution images from this camera allow scientists to search for signs of surface water and marks left by ancient shorelines.

✳ Visible and Infrared Mineralogical Mapping Spectrometer (OMEGA): This spectrometer precisely maps the composition of the Martian surface.

✳ Mars Radio Science Experiment (MARSIS): The long wavelengths of radar allow this detector to obtain information about the subsurface of Mars, as deep as 2.5 miles (4 km) below the surface. Scientists hope to find evidence for subsurface areas with this water or ice content.

✳ Planetary Fourier Spectrometer (PFS): This spectrometer measures the global composition and movement of the atmosphere.

✳ Ultraviolet and Infrared Atmospheric Spectrometer (SPICAM): This will look for traces of water and ozone in the atmosphere.

✳ Energetic Neutral Atoms Analyzer (ASPERA): This will study the way the atmosphere interacts with the wind of particles given off by the Sun.

The image shown on page 205 is how Mars appeared to the *Mars Global Surveyor* spacecraft on December 25, 2003, the day that *Beagle* and *Mars Express* reached the red planet. The large, dark region just left of center is Syrtis Major, a dark terrain known to astronomers for nearly four centuries before the first spacecraft went to Mars. Immediately to the right (east) of Syrtis Major is the somewhat circular plain, Isidis Planitia. *Beagle* arrived in Isidis Planitia only about 18 minutes before *Mars Global Surveyor* flew over the region. Relative to other global images of Mars taken over the past several years, the surface features were not as sharp and distinct because of considerable haze kicked up by large dust storms in the western and southern hemispheres during the previous two weeks. The fate of the *Beagle* lander is unknown.

Mars Express has produced as astonishing range of new scientific results, some of which lead to the new view that Mars went through an early wet stage, followed by an acidic, sulfurous stage, and ended with a dry cold stage. As of May 2009, the mission continues to send back images and surface data. NASA extended the mission until December 2012. *Mars Express* will continue to send back images and surface data until then.

The Mars Global Surveyor *took this image on December 25, 2003, the day that* Beagle 2 *and* Mars Express *reached the red planet. The large, dark region just left of center is Syrtis Major. Immediately to the right of Syrtis Major is the somewhat circular plain, Isidis Planitia.* Beagle 2 *arrived in Isidis Planitia only about 18 minutes before* Mars Global Surveyor *flew over the region and acquired a portion of this global view. A deep haze kicked up by dust storms obscures the details of the surface.* Beagle 2 *was successfully ejected from its mother craft,* Mars Express, *but nothing was heard of the mission after its ejection. (NASA/JPL/Malin Space Science Systems)*

Mars Exploration Rovers 2003
American

The Mars Exploration Rover (MER) mission successfully placed two robotic rovers on the Martian surface, in two distinctly different environments. The first launched on June 5, 2003, and landed on Mars January 4, 2004, and the second launched June 25, 2003, and landed January 25, 2004. Due to the need for maximum solar energy to power the rovers, they both landed near the Martian equator but on almost opposite sides of the planet. Over five years later they are still operating.

These rovers, *Spirit* and *Opportunity,* each carry five instruments to image the surface and analyze compositions:

✳ The Rock Abrasion Tool (RAT) is a powerful grinder, able to create a hole about two inches (45 mm) in diameter and 0.2 inches (5 mm) deep into a rock on the Martian surface. A rock sitting on the surface of Mars may become covered with dust and will weather, or change in chemical composition, from contact with the atmosphere. Once a fresh surface is exposed, scientists can examine the abraded area in detail using the rover's other science instruments

✳ The Miniature Thermal Emission Spectrometer (Mini-TES) is an infrared spectrometer that can determine the mineralogy of rocks and soils. One particular goal will be to search for minerals that were formed by the action of water, such as carbonates and clays. Mini-TES will also look at the atmosphere of Mars and gather data on temperature, water vapor, and the abundance of dust.

✳ The Mössbauer Spectrometer is an instrument that was specially designed to study iron-bearing minerals. Many of the minerals that formed rocks on Mars contain iron, and the soil is iron-rich. This instrument can determine the composition and abundance of these minerals with great accuracy. This ability can also help us understand the magnetic properties of surface materials. One Mössbauer measurement takes about 12 hours.

✳ The Alpha Particle X-ray Spectrometer (APXS) studies the alpha particles (emitted during radioactive decay) and X-rays emitted by rocks and soils to determine their composition. Knowing the elemental composition of Martian rocks provides scientists with information about the formation of the planet's crust, as well as any weathering that has taken place.

* Three sets of magnets that will collect airborne dust for analysis by the science instruments are carried by each rover. Magnetic minerals carried in dust grains may be freeze-dried remnants of the planet's watery past. One set of magnets will be carried by the Rock Abrasion Tool. A second set of two magnets is mounted on the front of the rover at an angle so that nonmagnetic particles will tend to fall off. These magnets can be reached by the Mössbauer and APXS instruments so they can analyze the magnetic particles. A third magnet is mounted on the top of the rover deck in view of the Pancam. This magnet is strong enough to deflect the paths of wind-carried, magnetic dust.

Scientists working on the rover missions have aligned their daily schedules with Mars. The Martian day is about 40 minutes longer than an Earth day, and so the scientists' days move forward by 40 minutes each 24 hours. The windows of the floors at the JPL in Pasadena, California, are covered with black shades to help the scientists with their ongoing time transition. To further complicate the problem, the two rovers have landed on opposite sides of the planet, and since they only operate in daylight the two science teams work on parallel schedules about 12 hours apart.

The rovers have been far more successful than anyone on their teams could ever have anticipated. They had modest primary science missions, just 90 Sols (Martian days) each. In late 2009, *Spirit* became stuck in soft soil, where it remains. It has since turned off all systems and altered hibernation, during which time the craft will charge its batteries. NASA is unsure when to expect this, so it has kept a vigilant watch. *Opportunity* is still active. Though the mission control space at the JPL had been planned for other missions starting several years ago, the ongoing exciting science and daily discoveries by these rovers has allowed the team to go on, on tighter budgets and with smaller teams, but still functioning.

These landers have captured mankind's imagination in a way that not even the most successful orbiting mission has succeeded in doing. Orbiters can remain in space sending data for years, while landers have expected lifetimes measured in weeks. Orbiters can measure gravity, composition, magnetic field, and topography over the entire planet, where landers can only measure and photograph what is right in front of them. Landers, to be sure, offer types of data that are not available from space-based sensors, but the impact they have on the public imagination must be based far more in their human-scale endeavors than on the relative quantity of the data they gather. Seeing the photos sent back from the surface allows us to imagine what it might be like to stand on another planet ourselves. The immediacy appeals to us in a personal way and makes each of us an explorer. Landers also pave the way to the final goal: sending a person to Mars.

Mars Reconnaissance Orbiter 2005
American

NASA launched the *Mars Reconnaissance Orbiter* on August 12, 2005, and the craft entered Mars orbit in March 2006. The mission is now in an extended phase and its main phase is considered a significant success. The *Reconnaissance Orbiter* imaged the surface of Mars with a resolution of just one to 1.5 feet (30 to 50 cm). It scanned underground layers for water and ice, identified small patches of surface minerals to determine their composition and origins, tracked changes in atmospheric water and dust, and checked global weather every day.

Probing below Mars's surface with penetrating radar, *Reconnaissance Orbiter*'s SHARAD ground-penetrating radar instrument was designed to discover whether the frozen water that NASA's *Mars Odyssey* spacecraft detected in the top yard or two of soil extends deeper, perhaps as accessible reservoirs of melted water. SHARAD can detect layers inside Mars to a depth of more than a half mile. One of its interesting results has come from examining areas in Mars that have gullies, thought by some scientists to be caused by groundwater discharge. SHARAD has found that there is no groundwater in these areas, and so the gullies must be formed by falling debris.

A more complete story of the *Mars Reconnaissance Orbiter*'s development, deployment, and science discoveries is found in the chapter on this mission.

Mars Phoenix Orbiter 2007
American

The first of the NASA Scout mission series, the *Phoenix* lander was designed for maximum science return on a relatively small budget as a part of an ongoing series that would ideally take advantage of every possible Mars launch date, which occurs every 26 months. The *Phoenix* lander was selected in the first Scout competition and launched August 4, 2007, landing in the north polar region of Mars on May 25, 2008. The lander is another key part of the NASA strategy for searching for signs of past or present microbial life in a series of Martian surface and near-surface environments. Its mission was to study the history of water and the potential for habitability on Mars.

The little lander is stationary and used a strong robotic arm to investigate the environment around its unusual northern polar landing site. The arm collected soil samples and delivered them to the instruments on the lander. There, miniature ovens and a mass spectrometer, an instrument that can measure the composition of material, awaited the samples. JPL called the collection of instruments a "chemistry lab in a box." Along with the ability to measure composition, the lander can look at samples with an atomic force microscope, capable of making images as small as 0.0000004 inches (10 nanometers), and with a series of cameras, capable of wide-field landscape pictures allowing the team to make images of samples up to 10^{12} times larger than those microscopic images.

Immediately upon landing, the mission scientists spotted interesting material in the soil beneath the lander. *Phoenix* blew away some dry surface soil upon landing and revealed what scientists thought might be ice. The robotic arm dug a trench in the frozen soil and hit a hard layer at about two inches depth (5 cm). After several failed attempts, during which the soil stuck inside the scoop or lost its water in a slow transfer to the lander, a soil sample with frozen water intact was delivered to the lander.

The main science mission lasted 90 Mars Sols, or 92 Earth days. At the end of July 2008, NASA announced that the *Phoenix* lander had identified water in a soil sample and operational funding for the mission was extended through September 30, 2008. Panoramic images from the camera boom show a landscape dominated by processes of frozen water, in agreement with observations from *Reconnaissance Orbiter*.

In the Martian winter, little sunlight reaches the poles and carbon dioxide frost was expected to block the lander's solar panels, causing the lander to gradually lose function as its power goes. Unlike the long-lived Mars Rover missions, *Phoenix* had a short planned lifetime that was absolutely limited by the lack of sunlight at the poles and could not be transcended by parking on slopes for hibernation or the other strategies that have helped the Rover missions survive. Additionally, images from the *Mars Reconnaissance Orbiter* show severe ice damage to the lander's solar panels. The last science transmission from *Phoenix* was on October 29, 2008, after 152 Martian days. On May 25, 2010, NASA officially ended the mission, after repeated unsuccessful attempts to contact the lander.

Future Missions

Though the lineup and launch dates change regularly with budgets, team changes, and reprioritizations, there remains a healthy schedule for Mars missions in the future. NASA continues to fill the 26-month cycle of possible Mars launches with planned missions.

A highly capable rover called *Mars Science Laboratory*, now officially named *Curiosity*, is deep in development for a 2011 launch opportunity, after delays from a planned 2009 launch. This mission is intended to be a long-range roving, long-duration science laboratory that will be a major leap in surface measurements and will pave the way for a future sample return mission. This capability will also demonstrate the technology for "smart landers" with accurate landing and hazard avoidance in order to reach what may be very

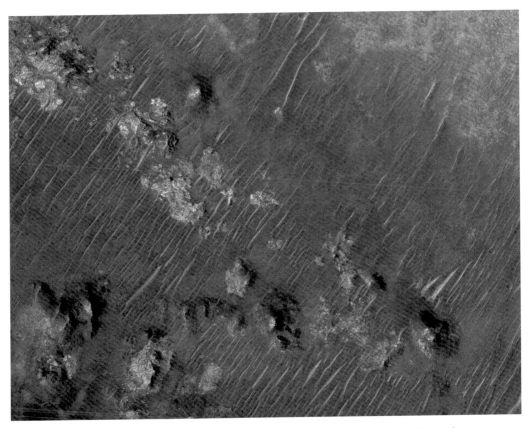

A portion of a trough in the Nili Fossae region of Mars is shown in enhanced color in this image from the HiRISE camera on Mars Reconnaissance Orbiter. The image was taken on March 24, 2007, as part of a campaign to examine more than two dozen candidate landing sites for the Mars Science Laboratory rover, which is scheduled for launch in 2011. The Nili Fossae region has one of the largest exposures of clay minerals discovered by orbital mapping spectrometers. Purple areas are basalt lava, blue-green areas are rich in the mineral pyroxene, and orange areas are clay, indicating reaction with water in the past. The image is about six-tenths of a mile wide. (NASA/JPL/ University of Arizona)

promising but difficult-to-reach scientific sites. The lander is expected to steer its own way past hazards to its landing site in much the way the space shuttle did on Earth.

The *Mars Science Laboratory* has a novel landing strategy as well. Its mass is too great to allow use of the successful balloon bounce landing method that the rovers used. Instead, a few minutes before touchdown, the craft will use its parachute and retro rockets to hang steady in the thin Martian atmosphere and will lower the lander to the Martian surface on a tether, much like a giant crane. This technology will allow the lander to choose a far smaller landing area than was required for bouncing or skidding landings.

The *Science Laboratory* is a large rover with six wheels and a mast for instruments. It will be able to collect rock and soil samples and bring them on board for chemical analysis, and it will be able to identify organic compounds and atmospheric gases associated with life. Its analysis abilities will paint a far more detailed picture of the water and carbon cycles on Mars and the geologic processes that formed the planet's surface.

Here an important link is also made with existing Mars missions, which will make planning and carrying out the ambitious lander mission possible. The *Mars Reconnaissance Orbiter*'s high-resolution instruments will help planners evaluate possible landing sites for the *Science Laboratory* both in terms of science potential for further discoveries and in terms of landing risks. The orbiter's communications capabilities will later provide a critical transmission relay for the surface mission.

Future Scout missions are lined up to 2011 and beyond. In the original 2006 round of Mars Scout mission proposals, two

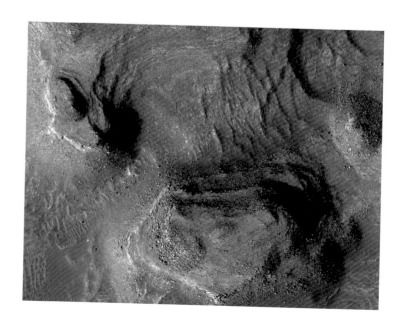

Mesas in the Nilosyrtis Mensae region of Mars appear in enhanced color in this image from the HiRISE camera on the Mars Reconnaissance Orbiter. This area, rich in clay minerals, is being considered as a landing site for the Mars Science Laboratory. (NASA/JPL/ University of Arizona)

2011 missions were chosen from a field of 26 proposals. These were the Mars Atmosphere and Volatile Evolution, or MAVEN, mission, and The Great Escape, or TGE mission. Both missions plan to measure the dynamics and evolution of the Martian atmosphere. The MAVEN mission is now scheduled for launch in 2013 and the TGE mission has been at least temporarily suspended. MAVEN, originally budgeted at $475 million, will certainly be rebudgeted to higher numbers with the delay.

International efforts to reach Mars are also continuing. The French NETLANDER mission, planned originally in 1998 for a 2009 launch, was canceled in 2003. Its ambitious goal to deploy the first extraterrestrial geophysical network created great interest in the scientific community. Robots and detailed high-resolution images make the news and stir up public support for missions, but to make the next important non-incremental step in science there would be little more useful than a geophysical network. NETLANDER would have sent four identical landers with seismometers (Mars certainly has earthquakes), ground-penetrating radar, electric field detectors, magnetometers, and instruments for measuring soil properties.

The data from such a network would have answered significant questions about the Martian interior and evolution, answers that cannot be obtained by single station missions. Such a networked mission is being planned for the Moon by NASA in agreement with a variety of international partners, and so there is hope that a similar mission to Mars might be attempted in the future. As a kind of successor to the NET-LANDER mission, the Finnish Meteorological Institute is planning MetNet, a series of landers on the Martian surface designed to measure atmospheric structure and dynamics.

The Russian Federation is planning a Mars mission called Phobos-Grunt, meaning Phobos soil. It is a robotic lander to be sent to Phobos and return a soil sample. Soil from Phobos would be scientifically interesting, as it would show the origin of the Moon and divulge information about surface processes. The mission had been scheduled for launch in 2011 but has been delayed. MetNet is planned for launch in several windows between 2011 and 2019.

The next round of Scout missions are tentatively scheduled for 2018. With the economic recession of 2009 continuing, there is little discussion of manned missions to Mars and even the manned missions to the Moon are being pushed off. There is still hope, however, for an eventual sample return from Mars. Sample return is a distant hope with no direct planning. In 2006, it was thought that during 2014 at the earliest a sample return mission would be launched, and the first fresh samples of rocks brought back from Mars. Now that date has been pushed back, but sample return is still a priority and a hope for many scientists. NASA is going through a major decadal planning period, run through committees at the National Research Council branch of the National Academy of Sciences. The outcomes of this long-term planning process will guide the choice of missions for the future.

Mars continues to be a main focus for planetary exploration. Both NASA and international space agencies have missions planned well into the future. A major goal of many of these missions is to determine whether conditions exist for life as it is understood on Earth, and to search for signs of life existing now, or from the past. Life on Mars, should it have existed, is hypothesized to have been single-celled organisms similar to the bacteria on Earth known as extremophiles. On Earth these organisms have adapted to life at temperatures near and even below freezing. As missions bring back more and more evidence for liquid water on Mars's surface in the past and for widely varying surface environments including acidic and basic briny regions, the search for life has more and more support.

Conclusions:
The Known and
the Unknown

Mars is the most-studied planet in current planetary science. Aside from the Moon, it is the closest and most hospitable of the terrestrial bodies, and fascinating especially for the possibility of finding evidence for life. The many missions that have succeeded in reaching Mars have sent back overwhelming quantities of data. This data has helped explain many questions about the planet (for example, there had to have been flowing liquid water on the surface in the past and there is seeping water today), it has created new questions that had not previously been asked (for example, what is the meaning of the many faint craters in the northern hemisphere? And how did Mars develop and then lose a strong magnetic dynamo?), and these missions also have yet to answer some crucial questions about the planet, for example:

1. **How did the Martian crust form?**
   On the Earth crust is constantly being destroyed and rebuilt through the actions of plate tectonics. There is no plate tectonics on Mars, and much of the crust is apparently ancient. Was it made as a result of magma ocean crystallization, when the solidified magma ocean overturned

into a stable density profile? If the northern plains are mainly andesitic, as remote sensing indicates, then how did half the planet come to be covered with a magma that on Earth requires water at depth in the mantle to form? The questions concerning the Martian crust are central to the formation and evolution of the planet itself.

2. **How did the planet heat originally, and why did it cool so fast and so early?**

Several lines of reasoning can tell scientists something about Mars's atmosphere in its early years. First, surface landforms show that Mars had flowing liquid surface water in the Noachian and perhaps in the Hesperian (corresponding to about 3.7 or about 3.2 billion years before the present). Surface temperatures therefore had to be above 32°F (0°C) for at least part of the year. Early impacts onto the planet, or the action of a freezing magma ocean, can be expected to have built up a steam-based atmosphere, but a very strong greenhouse effect is still needed to heat the planet to the point that water would remain liquid for hundreds of millions of years. On the other hand, based on isotopic examinations of the noble gases in the Martian atmosphere (argon and xenon primarily) the planet is thought to have lost 90 percent of its atmosphere to space by 1 billion years after formation (around three and a half billion years before present). With so little atmosphere left, how could the planet have remained warm enough on its surface to keep liquid water? Why did the atmosphere go so early in planetary evolution?

3. **Are Tharsis and the crustal dichotomy related? What about Tharsis and Hellas?**

Tharsis is an immense volcanic edifice near the Martian equator. On it reside three of Mars's largest shield volcanoes and just to its west, Olympus Mons, the largest volcano in the solar system. There are few other large volcanoes on the surface of the planet; the vast majority of volcanic activity is right at Tharsis. Tharsis forms one of several distinctive features on Mars; another is the crustal dichotomy, the line roughly around the planet's equator that marks the boundary between the southern highlands and the northern lowlands. The surface of Mars loses 2.5 miles (4 km) in elevation over a distance of about 180 miles (300 km) at the dichotomy. Tharsis straddles the dichotomy and extends to its north. Antipodal to Tharsis, to the south of the dichotomy on the opposite side of the planet, lies Hellas, the planet's largest impact crater. All three of these features are ancient and uniquely large. Could they be related? Though modeling results indicate that the formation of Hellas cannot cause convection in the mantle sufficient to create Tharsis, their near-exact *opposition* on the planet and apparent similar ages still link them in some scientist's minds. With no completely convincing hypothesis for the formation of either Tharsis or the dichotomy, perhaps a solution based on mantle convection will be introduced in coming years that will link the volcanism that created the dichotomy with that which created Tharsis.

4. **Was there ever life on Mars?**
   The discovery of life on Mars is the Holy Grail of modern planetary science. Encouraging information continues to form: The planet was once warm and wet, and even now is not too cold to support some of the unicellular life found on Earth; strange minerals and forms in one of the Martian meteorites continue to tantalize some scientists who think living creatures may have formed them; and Mars apparently has abundant sedimentary rocks in which to record any life that once existed. No clear indication of life has been found, but the search is just beginning. Though the discovery of life is perhaps the main driving force behind exploration of Mars, founding space mission exclusively on the search for life is a risky business. Any mission based on the search for a simple yes or no answer to a question is open to complete failure: If the answer appears to be no, how would the taxpayers of the nation feel about the (at minimum) hundreds of millions of dollars that were just spent? Broad questions are more successful, so the space mission can collect data from which many interesting

stories can be told. With luck, though, evidence for life on Mars may be found in the next few decades, and Earth creatures will no longer feel alone in the universe.

The large investment the American and other governments are making into Mars science makes this decade a great one to be a planetary scientist, full of exciting discoveries and opportunities. From Tharsis, Olympus Mons, and the crustal dichotomy to the unusual mantle compositions the Martian meteorites imply, Mars is a different world with different processes than the Earth has. From the valley networks and volcanic flows, though, Mars appears very much like Earth. How wonderful and strange to see such familiar proof of flowing water on another planet in an otherwise dry or frozen solar system. The complex similarities and differences between the planets make Mars an intriguing topic for research: Processes that apply to Earth will often apply to Mars, but with some interesting twists. Almost close enough to touch, close enough to consider sending human visitors to, Mars remains an alien planet with many unanswered questions.

APPENDIX 1:

Units and Measurements

FUNDAMENTAL UNITS

The system of measurements most commonly used in science is called both the SI (for Système International d'Unités) and the International System of Units (it is also sometimes called the MKS system). The SI system is based upon the metric units meter (abbreviated m), kilogram (kg), second (sec), kelvin (K), mole (mol), candela (cd), and ampere (A), used to measure length, time, mass, temperature, amount of a substance, light intensity, and electric current, respectively. This system was agreed upon in 1974 at an international general conference. There is another metric system, CGS, which stands for centimeter, gram, second; that system simply uses the hundredth of a meter (the centimeter) and the hundredth of the kilogram (the gram). The CGS system, formally introduced by the British Association for the Advancement of Science in 1874, is particularly useful to scientists making measurements of small quantities in laboratories, but it is less useful for space science. In this set, the SI system is used with the exception that temperatures will be presented in Celsius (C), instead of Kelvin. (The conversions between Celsius, Kelvin, and Fahrenheit temperatures are given below.) Often the standard unit of measure in the SI system, the meter, is too small when talking about the great distances in the solar system; kilometers (thousands of meters) or AU (astronomical units, defined below) will often be used instead of meters.

How is a unit defined? At one time a "meter" was defined as the length of a special metal ruler kept under strict conditions of temperature and humidity. That perfect meter could not be measured, however, without changing its temperature by opening the box, which would change its length, through thermal expansion or contraction. Today a meter is no longer defined according to a physical object; the only fundamental measurement that still is defined by a physical object is the kilogram. All of these units have had long and complex histories of attempts to define them. Some of the modern definitions, along with the use and abbreviation of each, are listed in the table here.

FUNDAMENTAL UNITS

Measurement	Unit	Symbol	Definition
length	meter	m	The meter is the distance traveled by light in a vacuum during 1/299,792,458 of a second.
time	second	sec	The second is defined as the period of time in which the oscillations of cesium atoms, under specified conditions, complete exactly 9,192,631,770 cycles. The length of a second was thought to be a constant before Einstein developed theories in physics that show that the closer to the speed of light an object is traveling, the slower time is for that object. For the velocities on Earth, time is quite accurately still considered a constant.
mass	kilogram	kg	The International Bureau of Weights and Measures keeps the world's standard kilogram in Paris, and that object is the definition of the kilogram.
temperature	kelvin	K	A degree in Kelvin (and Celsius) is 1/273.16 of the thermody- namic temperature of the triple point of water (the temper- ature at which, under one atmosphere pressure, water coexists as water vapor, liquid, and solid ice). In 1967, the General Conference on Weights and Measures defined this temperature as 273.16 kelvin.

Measurement	Unit	Symbol	Definition
amount of a substance	mole	mol	The mole is the amount of a substance that contains as many units as there are atoms in 0.012 kilogram of carbon 12 (that is, Avogadro's number, or 6.02205×10^{23}). The units may be atoms, molecules, ions, or other particles.
electric current	ampere	A	The ampere is that constant current which, if maintained in two straight parallel conductors of infinite length, of negligible circular cross section, and placed one meter apart in a vacuum, would produce between these conductors a force equal to 2×10^{-7} newtons per meter of length.
light intensity	candela	cd	The candela is the luminous intensity of a source that emits monochromatic radiation with a wavelength of 555.17 nm and that has a radiant intensity of 1/683 watt per steradian. Normal human eyes are more sensitive to the yellow-green light of this wavelength than to any other.

Mass and weight are often confused. Weight is proportional to the force of gravity: Your weight on Earth is about six times your weight on the Moon because Earth's gravity is about six times that of the Moon's. Mass, on the other hand, is a quantity of matter, measured independently of gravity. In fact, weight has different units from mass: Weight is actually measured as a force (newtons, in SI, or pounds, in the English system).

The table "Fundamental Units" lists the fundamental units of the SI system. These are units that need to be defined in order to make other measurements. For example, the meter and the second are fundamental units (they are not based on any other units). To measure velocity, use a derived unit, meters per second (m/sec), a combination of fundamental units. Later in this section there is a list of common derived units.

The systems of temperature are capitalized (Fahrenheit, Celsius, and Kelvin), but the units are not (degree and kelvin). Unit abbreviations are capitalized only when they are named after a person, such as K for Lord Kelvin, or A for André-Marie Ampère. The units themselves are always lowercase, even when named for a person: one newton, or one N. Throughout these tables a small dot indicates multiplication, as in N·m, which means a newton (N) times a meter (m). A space between the symbols can also be used to indicate multiplication, as in N m. When a small letter is placed in front of a symbol, it is a prefix meaning some multiplication factor. For example, J stands for the unit of energy called a joule, and a mJ indicates a millijoule, or 10^{-3} joules. The table of prefixes is given at the end of this section.

COMPARISONS AMONG KELVIN, CELSIUS, AND FAHRENHEIT

One kelvin represents the same temperature difference as 1°C, and the temperature in kelvins is always equal to 273.15 plus the temperature in degrees Celsius. The Celsius scale was designed around the behavior of water. The freezing point of water (at one atmosphere of pressure) was originally defined to be 0°C, while the boiling point is 100°C. The kelvin equals exactly 1.8°F.

To convert temperatures in the Fahrenheit scale to the Celsius scale, use the following equation, where F is degrees Fahrenheit, and C is degrees Celsius:

$$C = (F - 32)/1.8.$$

And to convert Celsius to Fahrenheit, use this equation:

$$F = 1.8C + 32.$$

To convert temperatures in the Celsius scale to the Kelvin scale, add 273.16. By convention, the degree symbol (°) is used for Celsius and Fahrenheit temperatures but not for temperatures given in Kelvin, for example, 0°C equals 273K.

What exactly is temperature? Qualitatively, it is a measurement of how hot something feels, and this definition is so easy to relate to that people seldom take it further. What is really happening in a substance as it gets hot or cold, and how does that change make temperature? When a fixed amount of energy is put into a substance, it heats up by an amount depending on what it is. The temperature of an object, then, has something to do with how the material responds to energy, and that response is called entropy. The entropy of a material (entropy is usually denoted S) is a measure of atomic wiggling and disorder of the atoms in the material. Formally, temperature is defined as

$$\frac{1}{T} = \left(\frac{dS}{dU} \right)_N ,$$

meaning one over temperature (the reciprocal of temperature) is defined as the change in entropy (dS, in differential notation) per change in energy (dU), for a given number of atoms (N). What this means in less technical terms is that temperature is a measure of how much heat it takes to increase the entropy (atomic wiggling and disorder) of a substance. Some materials get hotter with less energy, and others require more to reach the same temperature.

The theoretical lower limit of temperature is −459.67°F (−273.15°C, or 0K), known also as absolute zero. This is the temperature at which all atomic movement stops. The Prussian physicist Walther Nernst showed that it is impossible to actually reach absolute zero, though with laboratory methods using nuclear magnetization it is possible to reach 10 °K (0.000001K).

USEFUL MEASURES OF DISTANCE

A *kilometer* is a thousand meters (see the table "International System Prefixes"), and a *light-year* is the distance light travels in a vacuum during one year (exactly 299,792,458 m/sec, but commonly rounded to 300,000,000 m/sec). A light-year, therefore, is the distance that light can travel in one year, or:

$$299{,}792{,}458 \ m/sec \times 60 \ sec/min \times 60 \ min/hr \times$$
$$24 \ hr/day \times 365 \ days/yr = 9.4543 \ 10^{15} \ m/yr.$$

For shorter distances, some astronomers use light minutes and even light seconds. A light minute is 17,998,775 km, and a light second is 299,812.59 km. The nearest star to Earth, Proxima Centauri, is 4.2 light-years away from the Sun. The next, Rigil Centaurs, is 4.3 light-years away.

An *angstrom* (10^{-10}m) is a unit of length most commonly used in nuclear or particle physics. Its symbol is Å. The diameter of an atom is about one angstrom (though each element and isotope is slightly different).

An astronomical unit (AU) is a unit of distance used by astronomers to measure distances in the solar system. One astronomical unit equals the average distance from the center of the Earth to the center of the Sun. The currently accepted value, made standard in 1996, is 149,597,870,691 meters, plus or minus 30 meters.

One kilometer equals 0.62 miles, and one mile equals 1.61 kilometers.

The following table gives the most commonly used of the units derived from the fundamental units above (there are many more derived units not listed here because they have been developed for specific situations and are little-used elsewhere; for example, in the metric world, the curvature of a railroad track is measured with a unit called "degree of curvature," defined as the angle between two points in a curving track that are separated by a chord of 20 meters).

Though the units are given in alphabetical order for ease of reference, many can fit into one of several broad categories: dimensional units (angle, area, volume), material properties (density, viscosity, thermal expansivity), properties of motion (velocity, acceleration, angular velocity), electrical properties (frequency, electric charge, electric potential, resistance, inductance, electric field strength), magnetic properties (magnetic field strength, magnetic flux, magnetic flux density), and properties of radioactivity (amount of radioactivity and effect of radioactivity).

DERIVED UNITS

Measurement	Unit symbol (derivation)	Comments
acceleration	unnamed (m/sec^2)	
angle	radian rad (m/m)	One radian is the angle centered in a circle that includes an arc of length equal to the radius. Since the circumference equals two pi times the radius, one radian equals 1/(2 pi) of the circle, or approximately 57.296°.
	steradian sr (m^2/m^2)	The steradian is a unit of solid angle. There are four pi steradians in a sphere. Thus one steradian equals about 0.079577 sphere, or about 3282.806 square degrees.
angular velocity	unnamed (rad/sec)	
area	unnamed (m^2)	
density	unnamed (kg/m^3)	Density is mass per volume. Lead is dense, styrofoam is not. Water has a density of one gram per cubic centimeter or 1,000 kilograms per cubic meter.
electric charge or electric flux	coulomb C (A·sec)	One coulomb is the amount of charge accumulated in one second by a current of one ampere. One coulomb is also the amount of charge on 6.241506×10^{18} electrons.
electric field strength	unnamed [(kg·m)/ (sec^3·A) × V/m]	Electric field strength is a measure of the intensity of an electric field at a particular location. A field strength of one V/m represents a potential difference of one volt between points separated by one meter.
electric potential, or electromotive force (often called voltage)	volt V [(kg·m^2)/ (sec^3·A) = J/C = W/A]	Voltage is an expression of the potential difference in charge between two points in an electrical field. Electric potential is defined as the amount of potential energy present per unit of charge. One volt is a potential of one joule per coulomb of charge. The greater the voltage, the greater the flow of electrical current.

(continues)

DERIVED UNITS (continued)

Measurement	Unit symbol (derivation)	Comments
energy, work, or heat	joule J [N·m (=kg·m^2/sec^2)]	
	electron volt eV	The electron volt, being so much smaller than the joule (one eV = 1.6 × 10^{-17} J), is useful for describing small systems.
force	newton N (kg·m/sec^2)	This unit is the equivalent to the pound in the English system, since the pound is a measure of force and not mass.
frequency	hertz Hz (cycles/sec)	Frequency is related to wavelength as follows: kilohertz × wavelength in meters = 300,000.
inductance	henry H (Wb/A)	Inductance is the amount of magnetic flux a material pro- duces for a given current of electricity. Metal wire with an electric current passing through it creates a magnetic field; different types of metal make magnetic fields with different strengths and therefore have different inductances.
magnetic field strength	unnamed (A/m)	Magnetic field strength is the force that a magnetic field exerts on a theoretical unit magnetic pole.
magnetic flux	weber Wb [(kg·m^2)/ (sec^2·A) = V·sec]	The magnetic flux across a perpendicular surface is the product of the magnetic flux density, in teslas, and the surface area, in square meters.
magnetic flux density	tesla T [kg/(sec^2·A) = Wb/m^2]	A magnetic field of one tesla is strong: The strongest artificial fields made in laboratories are about 20 teslas, and the Earth's magnetic flux density, at its surface, is about 50 microteslas (μT). Planetary magnetic fields are sometimes measured in gammas, which are nanoteslas (10^{-9} teslas).
momentum, or impulse	unnamed [N·sec (= kg m/sec)]	Momentum is a measure of moving mass: how much mass and how fast it is moving.

Measurement	Unit symbol (derivation)	Comments
power	watt W [J/s (= $(kg\ m^2)/sec^3$)]	Power is the rate at which energy is spent. Power can be mechanical (as in horsepower) or electrical (a watt is produced by a current of one ampere flowing through an electric potential of one volt).
pressure, or stress	pascal Pa (N/m^2)	The high pressures inside planets are often measured in gigapascals (10^9 pascals), abbreviated GPa. ~10,000 atm = one GPa.
	atmosphere atm	The atmosphere is a handy unit because one atmosphere is approximately the pressure felt from the air at sea level on Earth; one standard atm = 101,325 Pa; one metric atm = 98,066 Pa; one atm ~ one bar.
radiation per unit mass receiving it	gray (J/kg)	The amount of radiation energy absorbed per kilogram of mass. One gray = 100 rads, an older unit.
radiation (effect of)	sievert Sv	This unit is meant to make comparable the biological effects of different doses and types of radiation. It is the energy of radiation received per kilogram, in grays, multiplied by a factor that takes into consideration the damage done by the particular type of radiation.
radioactivity (amount)	becquerel Bq	One atomic decay per second
	curie Ci	The curie is the older unit of measure but is still frequently seen. One Ci = 3.7×10^{10} Bq.
resistance	ohm Ω (V/A)	Resistance is a material's unwillingness to pass electric current. Materials with high resistance become hot rather than allowing the current to pass and can make excellent heaters.
thermal expansivity	unnamed (/°)	This unit is per degree, measuring the change in volume of a substance with the rise in temperature.

(continues)

DERIVED UNITS *(continued)*		
Measurement	**Unit symbol (derivation)**	**Comments**
vacuum	torr	Vacuum is atmospheric pressure below one atm (one torr = 1/760 atm). Given a pool of mercury with a glass tube standing in it, one torr of pressure on the pool will press the mercury one millimeter up into the tube, where one standard atmosphere will push up 760 millimeters of mercury.
velocity	unnamed (m/sec)	
viscosity	unnamed [Pa·sec (= kg/(m·sec)]	Viscosity is a measure of resistance to flow. If a force of one newton is needed to move one square meter of the liquid or gas relative to a second layer one meter away at a speed of one meter per second, then its viscosity is one Pa·s, often simply written Pas or Pas. The cgs unit for viscosity is the poise, equal to 0.1Pa·s.
volume	cubic meter (m³)	

DEFINITIONS FOR ELECTRICITY AND MAGNETISM

When two objects in each other's vicinity have different electrical charges, an *electric field* exists between them. An electric field also forms around any single object that is electrically charged with respect to its environment. An object is negatively charged (–) if it has an excess of electrons relative to its surroundings. An object is positively charged (+) if it is deficient in electrons with respect to its surroundings.

An electric field has an effect on other charged objects in the vicinity. The field strength at a particular distance from an object is directly proportional to the electric charge of that object, in coulombs. The field strength is inversely proportional to the distance from a charged object.

Flux is the rate (per unit of time) in which something flowing crosses a surface perpendicular to the direction of flow.

An alternative expression for the intensity of an electric field is *electric flux density*. This refers to the number of lines of electric flux passing at right angles through a given surface area, usually one meter squared (1 m²). Electric flux density, like electric field strength, is directly proportional to the

INTERNATIONAL SYSTEM PREFIXES

SI prefix	Symbol	Multiplying factor
exa-	E	10^{18} = 1,000,000,000,000,000,000
peta-	P	10^{15} = 1,000,000,000,000,000
tera-	T	10^{12} = 1,000,000,000,000
giga-	G	10^{9} = 1,000,000,000
mega-	M	10^{6} = 1,000,000
kilo-	k	10^{3} = 1,000
hecto-	h	10^{2} = 100
deca-	da	10 = 10
deci-	d	10^{-1} = 0.1
centi-	c	10^{-2} = 0.01
milli-	m	10^{-3} = 0.001
micro-	μ or u	10^{-6} = 0.000,001
nano-	n	10^{-9} = 0.000,000,001
pico-	p	10^{-12} = 0.000,000,000,001
femto-	f	10^{-15} = 0.000,000,000,000,001
atto-	a	10^{-18} = 0.000,000,000,000,000,001

A note on nonmetric prefixes: In the United States, the word billion means the number 1,000,000,000, or 10^{9}. In most countries of Europe and Latin America, this number is called "one milliard" or "one thousand million," and "billion" means the number 1,000,000,000,000, or 10^{12}, which is what Americans call a "trillion." In this set, a billion is 10^{9}.

NAMES FOR LARGE NUMBERS

Number	American	European	SI prefix
10^9	billion	milliard	giga-
10^{12}	trillion	billion	tera-
10^{15}	quadrillion	billiard	peta-
10^{18}	quintillion	trillion	exa-
10^{21}	sextillion	trilliard	zetta-
10^{24}	septillion	quadrillion	yotta-
10^{27}	octillion	quadrilliard	
10^{30}	nonillion	quintillion	
10^{33}	decillion	quintilliard	
10^{36}	undecillion	sextillion	
10^{39}	duodecillion	sextilliard	
10^{42}	tredecillion	septillion	
10^{45}	quattuordecillion	septilliard	

This naming system is designed to expand indefinitely by factors of powers of three. Then, there is also the googol, the number 10^{100} (one followed by 100 zeroes). The googol was invented for fun by the eight-year-old nephew of the American mathematician Edward Kasner. The googolplex is 10^{googol}, or one followed by a googol of zeroes. Both it and the googol are numbers larger than the total number of atoms in the universe, thought to be about 10^{80}.

charge on the object. But flux density diminishes with distance according to the inverse-square law because it is specified in terms of a surface area (per meter squared) rather than a linear displacement (per meter).

A *magnetic field* is generated when electric charge carriers such as electrons move through space or within an electrical conductor. The geometric shapes of the magnetic flux lines produced by moving charge carriers (electric current) are similar to the shapes of the flux lines in an electrostatic field. But there are differences in the ways electrostatic and magnetic fields interact with the environment.

Electrostatic flux is impeded or blocked by metallic objects. *Magnetic flux* passes through most metals with little or no effect, with certain exceptions, notably iron and nickel. These two metals, and alloys and mixtures containing them, are known as ferromagnetic materials because they concentrate magnetic lines of flux.

Magnetic flux density and *magnetic force* are related to *magnetic field strength*. In general, the magnetic field strength diminishes with increasing distance from the axis of a magnetic dipole in which the flux field is stable. The function defining the rate at which this field-strength decrease occurs depends on the geometry of the magnetic lines of flux (the shape of the flux field).

PREFIXES

Adding a prefix to the name of that unit forms a multiple of a unit in the International System (see the table "International System Prefixes"). The prefixes change the magnitude of the unit by orders of 10 from 10^{18} to 10^{-18}.

Very small concentrations of chemicals are also measured in parts per million (ppm) or parts per billion (ppb), which mean just what they sound like: If there are four parts per million of lead in a rock (4 ppm), then out of every million atoms in that rock, on average four of them will be lead.

APPENDIX 2:

Light, Wavelength, and Radiation

Electromagnetic radiation is energy given off by matter, traveling in the form of waves or particles. Electromagnetic energy exists in a wide range of energy values, of which visible light is one small part of the total spectrum. The source of radiation may be the hot and therefore highly energized atoms of the Sun, pouring out radiation across a wide range of energy values, including of course visible light, and they may also be unstable (radioactive) elements giving off radiation as they decay.

Radiation is called "electromagnetic" because it moves as interlocked waves of electrical and magnetic fields. A wave is a disturbance traveling through space, transferring energy from one point to the next. In a vacuum, all electromagnetic radiation travels at the speed of light, 983,319,262 feet per second (299,792,458 m/sec, often approximated as 300,000,000 m/sec). Depending on the type of radiation, the waves have different wavelengths, energies, and frequencies (see the following figure). The wavelength is the distance between individual waves, from one peak to another. The frequency is the number of waves that pass a stationary point each second. Notice in the graphic how the wave undulates up and down from peaks to valleys to peaks. The time from one peak to the next peak is called one cycle. A single unit of frequency is equal to one cycle per second. Scientists refer to a single cycle as one hertz, which commemorates 19th-century German physicist Heinrich Hertz, whose discovery of electromagnetic waves led to the development of radio. The frequency of a wave is related to its energy: The higher the frequency of a wave, the higher its energy, though its speed in a vacuum does not change.

The smallest wavelength, highest energy and frequency electromagnetic waves are cosmic rays, then as wavelength increases and energy and frequency decrease, come gamma rays, then X-rays, then ultraviolet light, then visible light (moving from violet through indigo, blue, green, yellow, orange, and red), then infrared (divided into near, meaning near to visible, mid-, and far infrared), then microwaves, and then radio waves, which have the longest wavelengths and the lowest energy and frequency. The electromagnetic spectrum is shown in the accompanying figure and table.

As a wave travels and vibrates up and down with its characteristic wavelength, it can be imagined as vibrating up and down in a single plane, such as the plane of this sheet of paper in the case of the simple example in the figure here showing polarization. In nature, some waves change their polarization constantly so that their polarization sweeps through all angles, and they are said to be circularly polarized. In ordinary visible light, the waves are vibrating up and down in numerous random planes. Light can be shone through a special filter called a polarizing filter that blocks out all the light except that polarized in a certain direction, and the light that shines out the other side of the filter is then called polarized light.

Each electromagnetic wave has a measurable wavelength and frequency.

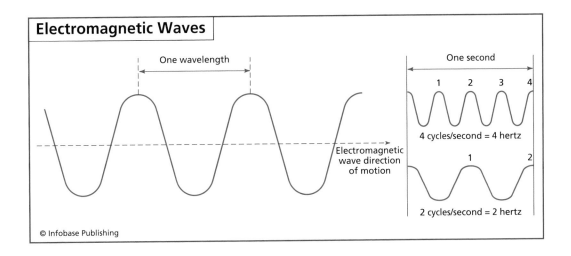

Electromagnetic Waves

One wavelength

One second

1 2 3 4

4 cycles/second = 4 hertz

Electromagnetic wave direction of motion

1   2

2 cycles/second = 2 hertz

© Infobase Publishing

Polarization is important in wireless communications systems such as radios, cell phones, and non-cable television. The orientation of the transmitting antenna creates the polarization of the radio waves transmitted by that antenna: A vertical antenna emits vertically polarized waves, and a horizontal antenna emits horizontally polarized waves. Similarly, a horizontal antenna is best at receiving horizontally polarized waves and a vertical antenna at vertically polarized waves.

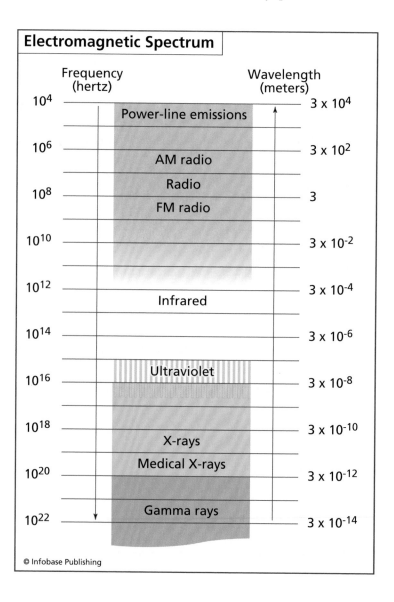

The electromagnetic spectrum ranges from cosmic rays at the shortest wavelengths to radiowaves at the longest wavelengths.

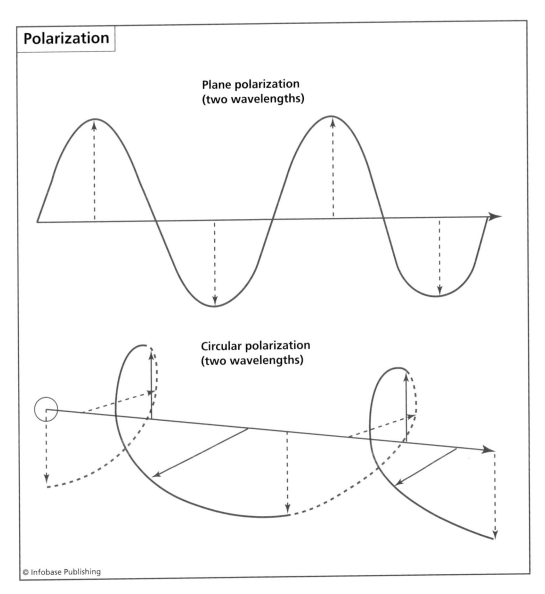

Polarization

Plane polarization
(two wavelengths)

Circular polarization
(two wavelengths)

© Infobase Publishing

The best communications are obtained when the source and receiver antennas have the same polarization. This is why, when trying to adjust television antennas to get a better signal, having the two antennae at right angles to each other can maximize the chances of receiving a signal.

The human eye stops being able to detect radiation at wavelengths between 3,000 and 4,000 angstroms, which is deep violet—also the rough limit on transmissions through

Waves can be thought of as plane or circularly polarized.

the atmosphere (see the table "Wavelengths and Frequencies of Visible Light"). (Three thousand to 4,000 angstroms is the same as 300–400 nm because an angstrom is 10^{-9} m, while the prefix nano- or n means 10^{-10}; for more, see appendix 1, "Units and Measurements.") Of visible light, the colors red, orange, yellow, green, blue, indigo, and violet are listed in order from longest wavelength and lowest energy to shortest wavelength and highest energy. Sir Isaac Newton, the spectacular English physicist and mathematician, first found that a glass prism split sunlight into a rainbow of colors. He named this a "spectrum," after the Latin word for ghost.

If visible light strikes molecules of gas as it passes through the atmosphere, it may get absorbed as energy by the molecule. After a short amount of time, the molecule releases the light, most probably in a different direction. The color that is radiated is the same color that was absorbed. All the colors of visible light can be absorbed by atmospheric molecules, but the higher energy blue light is absorbed more often than the lower energy red light. This process is called Rayleigh scattering (named after Lord John Rayleigh, an English physicist who first described it in the 1870s).

The blue color of the sky is due to Rayleigh scattering. As light moves through the atmosphere, most of the longer wave-

WAVELENGTHS AND FREQUENCIES OF VISIBLE LIGHT

Visible light color	Wavelength (in Å, angstroms)	Frequency (times 10^{14} Hz)
violet	4,000–4,600	7.5–6.5
indigo	4,600–4,750	6.5–6.3
blue	4,750–4,900	6.3–6.1
green	4,900–5,650	6.1–5.3
yellow	5,650–5,750	5.3–5.2
orange	5,750–6,000	5.2–5.0
red	6,000–8,000	5.0–3.7

WAVELENGTHS AND FREQUENCIES OF THE ELECTROMAGNETIC SPECTRUM

Energy	Frequency in hertz (Hz)	Wavelength in meters
cosmic rays	everything higher in energy than gamma rays	everything lower in wavelength than gamma rays
gamma rays	10^{20} to 10^{24}	less than 10^{-12} m
X-rays	10^{17} to 10^{20}	1 nm to 1 pm
ultraviolet	10^{15} to 10^{17}	400 nm to 1 nm
visible	4×10^{14} to 7.5×10^{14}	750 nm to 400 nm
near-infrared	1×10^{14} to 4×10^{14}	2.5 µm to 750 nm
infrared	10^{13} to 10^{14}	25 µm to 2.5 µm
microwaves	3×10^{11} to 10^{13}	1 mm to 25 µm
radio waves	less than 3×10^{11}	more than 1 mm

lengths pass straight through: The air affects little of the red, orange, and yellow light. The gas molecules absorb much of the shorter wavelength blue light. The absorbed blue light is then radiated in different directions and is scattered all around the sky. Whichever direction you look, some of this scattered blue light reaches you. Since you see the blue light from everywhere overhead, the sky looks blue. Note also that there is a very different kind of scattering, in which the light is simply bounced off larger objects like pieces of dust and water droplets, rather than being absorbed by a molecule of gas in the atmosphere and then reemitted. This bouncing kind of scattering is responsible for red sunrises and sunsets.

Until the end of the 18th century, people thought that visible light was the only kind of light. The amazing amateur astronomer Frederick William Herschel (the discoverer of Uranus) discovered the first non-visible light, the infrared. He thought that each color of visible light had a different temperature and devised an experiment to measure the temperature of each color of light. The temperatures went up as the colors

COMMON USES FOR RADIO WAVES

User	Approximate frequency
AM radio	0.535×10^6 to 1.7×10^6 Hz
baby monitors	49×10^6 Hz
cordless phones	49×10^6 Hz
	900×10^6 Hz
	$2,400 \times 10^6$ Hz
television channels 2 through 6	54×10^6 to 88×10^6 Hz
radio-controlled planes	72×10^6 Hz
radio-controlled cars	75×10^6 Hz
FM radio	88×10^6 to 108×10^6 Hz
television channels 7 through 13	174×10^6 to 220×10^6 Hz
wildlife tracking collars	215×10^6 Hz
cell phones	800×10^6 Hz
	$2,400 \times 10^6$ Hz
air traffic control radar	960×10^6 Hz
	$1,215 \times 10^6$ Hz
global positioning systems	$1,227 \times 10^6$ Hz
	$1,575 \times 10^6$ Hz
deep space radio	$2,300 \times 10^6$ Hz

progressed from violet through red, and then Herschel decided to measure past red, where he found the highest temperature yet. This was the first demonstration that there was a kind of radiation that could not be seen by the human eye. Herschel originally named this range of radiation "calorific rays," but the name was later changed to infrared, meaning "below red."

Infrared radiation has become an important way of sensing solar system objects and is also used in night-vision goggles and various other practical purposes.

At lower energies and longer wavelengths than the visible and infrared, microwaves are commonly used to transmit energy to food in microwave ovens, as well as for some communications, though radio waves are more common in this use. There is a wide range of frequencies in the radio spectrum, and they are used in many ways, as shown in the table "Common Uses for Radio Waves," including television, radio, and cell phone transmissions. Note that the frequency units are given in terms of 10^6 Hz, without correcting for each coefficient's additional factors of 10. This is because 10^6 Hz corresponds to the unit of megahertz (MHz), which is a commonly used unit of frequency.

Cosmic rays, gamma rays, and X-rays, the three highest-energy radiations, are known as ionizing radiation because they contain enough energy that, when they hit an atom, they may knock an electron off of it or otherwise change the atom's weight or structure. These ionizing radiations, then, are particularly dangerous to living things; for example, they can damage DNA molecules (though good use is made of them as well, to see into bodies with X-rays and to kill cancer cells with gamma rays). Luckily the atmosphere stops most ionizing radiation, but not all of it. Cosmic rays created by the Sun in solar flares, or sent off as a part of the solar wind, are relatively low energy. There are far more energetic cosmic rays, though, that come from distant stars through interstellar space. These are energetic enough to penetrate into an asteroid as deeply as a meter and can often make it through the atmosphere.

When an atom of a radioisotope decays, it gives off some of its excess energy as radiation in the form of X-rays, gamma rays, or fast-moving subatomic particles: alpha particles (two protons and two neutrons, bound together as an atomic *nucleus*), or beta particles (fast-moving electrons), or a combination of two or more of these products. If it decays with emission of an alpha or beta particle, it becomes a new element. These decay products can be described as gamma, beta, and alpha radiation.

RADIOACTIVITY OF SELECTED OBJECTS AND MATERIALS

Object or material	Radioactivity
1 adult human (100 Bq/kg)	7,000 Bq
1 kg coffee	1,000 Bq
1 kg high-phosphate fertilizer	5,000 Bq
1 household smoke detector (with the element americium)	30,000 Bq
radioisotope source for cancer therapy	100 million million Bq
1 kg 50-year-old vitrified high-level nuclear waste	10 million million Bq
1 kg uranium ore (Canadian ore, 15% uranium)	25 million Bq
1 kg uranium ore (Australian ore, 0.3% uranium)	500,000 Bq
1 kg granite	1,000 Bq

By decaying, the atom is progressing in one or more steps toward a stable state where it is no longer radioactive.

The X-rays and gamma rays from decaying atoms are identical to those from other natural sources. Like other ionizing radiation, they can damage living tissue but can be blocked by lead sheets or by thick concrete. Alpha particles are much larger and can be blocked more quickly by other material; a sheet of paper or the outer layer of skin on your hand will stop them. If the atom that produces them is taken inside the body, however, such as when a person breathes in radon gas, the alpha particle can do damage to the lungs. Beta particles are more energetic and smaller and can penetrate a couple of centimeters into a person's body.

But why can both radioactive decay that is formed of sub-atomic particles and heat that travels as a wave of energy be considered radiation? One of Albert Einstein's great discoveries is called the photoelectric effect: Subatomic particles can all behave as either a wave or a particle. The smaller the particle, the more wavelike it is. The best example of this is light itself, which behaves almost entirely as a wave, but there is the particle equivalent for light, the massless photon. Even alpha particles, the largest decay product discussed here, can act like a wave, though their wavelike properties are much harder to detect.

The amount of radioactive material is given in becquerel (Bq), a measure that enables us to compare the typical radioactivity of some natural and other materials. A becquerel is one atomic decay per second. Radioactivity is still sometimes measured using a unit called a Curie; a Becquerel is 27×10^{-12} Curies. There are materials made mainly of radioactive elements, like uranium, but most materials are made mainly of stable atoms. Even materials made mainly of stable atoms, however, almost always have trace amounts of radioactive elements in them, and so even common objects give off some level of radiation, as shown in the following table.

Background radiation is all around us all the time. Naturally occurring radioactive elements are more common in some kinds of rocks than others; for example, *granite* carries more radioactive elements than does sandstone; therefore a person working in a bank built of granite will receive more radiation than someone who works in a wooden building. Similarly, the atmosphere absorbs cosmic rays, but the higher the elevation, the more cosmic-ray exposure there is. A person living in Denver or in the mountains of Tibet is exposed to more cosmic rays than someone living in Boston or in the Netherlands.

APPENDIX 3:

A list of All Known Moons

Though Mercury and Venus have no moons, the other planets in the solar system have at least one. Some moons, such as Earth's Moon and Jupiter's Galileans satellites, are thought to have formed at the same time as their accompanying planet. Many other moons appear simply to be captured asteroids; for at least half of Jupiter's moons, this seems to be the case. These small, irregular moons are difficult to detect from Earth, and so the lists given in the table below must be considered works in progress for the gas giant planets. More moons will certainly be discovered with longer observation and better instrumentation.

KNOWN MOONS OF ALL PLANETS

Earth	Mars	Jupiter	Saturn	Uranus	Neptune	Pluto
1	2	63	62	27	13	3
1. Moon	1. Phobos	1. Metis	1. S/2009 S1	1. Cordelia	1. Naiad	1. Charon
	2. Diemos	2. Adrastea	2. Pan	2. Ophelia	2. Thalassa	2. Nix (P1)
		3. Amalthea	3. Daphnis	3. Bianca	3. Despina	3. Hydra
		4. Thebe	4. Atlas	4. Cressida	4. Galatea	(P2)
		5. Io	5. Prometheus	5. Desdemona	5. Larissa	
		6. Europa	6. Pandora	6. Juliet	6. Proteus	
		7. Ganymede	7. Epimetheus	7. Portia	7. Triton	
		8. Callisto	8. Janus	8. Rosalind	8. Nereid	
		9. Themisto	9. Aegaeon	9. Cupid (2003	9. Halimede	
		10. Leda	10. Mimas	U2)	(S/2002	
		11. Himalia	11. Methone	10. Belinda	N1)	

Earth	Mars	Jupiter	Saturn	Uranus	Neptune	Pluto
		12. Lysithea	12. Anthe	11. Perdita	10. Sao	
		13. Elara	13. Pallene	(1986 U10)	(S/2002 N2)	
		14. S/2000 J11	14. Enceladus	12. Puck	11. Laomedeia	
		15. Carpo	15. Telesto	13. Mab (2003	(S/2002	
		(S/2003	16. Tethys	U1)	N3)	
		J20)	17. Calypso	14. Miranda	12. Psamathe	
		16. S/2003	18. Dione	15. Ariel	(S/2003	
		J12	19. Helene	16. Umbriel	N1)	
		17. Euporie	20. Polydeuces	17. Titania	13. Neso	
		18. S/2003 J3	21. Rhea	18. Oberon	(S/2002	
		19. S/2003 J18	22. Titan	19. Francisco	N4)	
		20. Orthosie	23. Hyperion	(2001 U3)		
		21. Euanthe	24. Iapetus	20. Caliban		
		22. Harpalyke	25. Kiviuq	21. Stephano		
		23. Praxidike	26. Ijiraq	22. Trinculo		
		24. Thyone	27. Phoebe	23. Sycorax		
		25. S/2003	28. Paaliaq	24. Margaret		
		J16	29. Skathi	(2003 U3)		
		26. Mneme	30. Albiorix	25. Prospero		
		(S/2003	31. S/2007	26. Setebos		
		J21)	S2	27. Ferdinand		
		27. Iocaste	32. Bebhionn	(2001 U2)		
		28. Helike	33. Erriapo			
		(S/2003 J6)	34. Siarnaq			
		29. Hermippe	35. Skoll			
		30. Thelxinoe	36. Tarvos			
		(S/2003	37. Tarqeq			
		J22)	38. Greip			
		31. Ananke	39. Hyrrokkin			
		32. S/2003	40. S/2004			
		J15	S13			
		33. Eurydome	41. S/2004			
		34. S/2003	S17			
		J17	42. Mundilfari			
		35. Pasithee	43. Jarnsaxa			
		36. S/2003	44. S/2006			
		J10	S1			

(continues)

KNOWN MOONS OF ALL PLANETS *(continued)*

Earth	Mars	Jupiter	Saturn	Uranus	Neptune	Pluto
		37. Chaldene	45. Narvi			
		38. Isonoe	46. Bergelmir			
		39. Erinome	47. Suttungr			
		40. Kale	48. S/2004			
		41. Aitne	S12			
		42. Taygete	49. S/2004			
		43. Kallichore	S7			
		(S/2003	50. Hati			
		J11)	51. Bestla			
		44. Eukelade	52. Farbauti			
		(S/2003 J1)	53. Thrymyr			
		45. Arche	54. S/2007			
		(S/2002 J1)	S3			
		46. S/2003 J9	55. Aegir			
		47. Carme	56. S/2006			
		48. Kalyke	S3			
		49. Sponde	57. Kari			
		50. Magaclite	58. Fenrir			
		51. S/2003 J5	59. Surtur			
		52. S/2003	60. Ymir			
		J19	61. Loge			
		53. S/2003	62. Fornjot			
		J23				
		54. Hegemone				
		(S/2003 J8)				
		55. Pasiphae				
		56. Cyllene				
		(S/2003				
		J13)				
		57. S/2003 J4				
		58. Sinope				
		59. Aoede				
		(S/2003 J7)				
		60. Autonoe				
		61. Calirrhoe				
		62. Kore				
		(S/2003				
		J14)				
		63. S/2003 J2				

Glossary

accretion The accumulation of celestial gas, dust, or smaller bodies by gravitational attraction into a larger body, such as a planet or an asteroid

achondite A stony (silicate-based) meteorite that contains no chondrules; these originate in differentiated bodies and may be mantle material or lavas (see also chondrite and iron meteorite)

albedo The light reflected by an object as a fraction of the light shining on an object; mirrors have high albedo, while charcoal has low albedo

anorthite A calcium-rich plagioclase mineral with compositional formula $CaAl_2Si_2O_8$, significant for making up the majority of the rock anorthosite in the crust of the Moon

anticyclone An area of increased atmospheric pressure relative to the surrounding pressure field in the atmosphere, resulting in circular flow in a clockwise direction north of the equator and in a counterclockwise direction to the south

aphelion A distance; the farthest from the Sun an object travels in its orbit

apogee As for aphelion but for any orbital system (not confined to the Sun)

apparent magnitude The brightness of a celestial object as it would appear from a given distance—the lower the number, the brighter the object

atom The smallest quantity of an element that can take part in a chemical reaction; consists of a nucleus of protons and neutrons, surrounded by a cloud of electrons; each atom is about 10^{-10} meters in diameter, or one angstrom

atomic number   The number of protons in an atom's nucleus

AU   An AU is an astronomical unit, defined as the distance from the Sun to the Earth; approximately 93 million miles, or 150 million kilometers. For more information, refer to the UNITS AND MEASUREMENTS appendix

basalt   A generally dark-colored extrusive igneous rock most commonly created by melting a planet's mantle; its low silica content indicates that it has not been significantly altered on its passage to the planet's surface

bolide   An object falling into a planet's atmosphere, when a specific identification as a comet or asteroid cannot be made

bow shock   The area of compression in a flowing fluid when it strikes an object or another fluid flowing at another rate; for example, the bow of a boat and the water, or the magnetic field of a planet and the flowing solar wind

breccia   Material that has been shattered from grinding, as in a fault, or from impact, as by meteorites or other solar system bodies

CAIs   Calcium-aluminum inclusions, small spheres of mineral grains found in chondritic meteorites and thought to be the first solids that formed in the protoplanetary disk

calcium-aluminum inclusion   See **CAIs**

chondrite   A class of meteorite thought to contain the most primitive material left from the solar nebula; named after their glassy, super-primitive inclusions called chondrules

chondrule   Rounded, glassy, and crystalline bodies incorporated into the more primitive of meteorites; thought to be the condensed droplets of the earliest solar system materials

CI chondrite   The class on chondrite meteorites with compositions most like the Sun, and therefore thought to be the oldest and least altered material in the solar system

clinopyroxene   A common mineral in the mantle and igneous rocks, with compositional formula $((Ca,Mg,Fe,Al)_2(Si,Al)_2O_6)$

conjunction When the Sun is between the Earth and the planet or another body in question

convection Material circulation upward and downward in a gravity field caused by horizontal gradients in density; an example is the hot, less dense bubbles that form at the bottom of a pot, rise, and are replaced by cooler, denser sinking material

core The innermost material within a differentiated body; in a rocky planet this consists of iron-nickel metal, and in a gas planet this consists of the rocky innermost solids

Coriolis force The effect of movement on a rotating sphere; movement in the Northern Hemisphere curves to the right, while movement in the Southern Hemisphere curves to the left

craton The ancient, stable interior cores of the Earth's continents

crust The outermost layer of most differentiated bodies, often consisting of the least dense products of volcanic events or other buoyant material

cryovolcanism Non-silicate materials erupted from icy and gassy bodies in the cold outer solar system; for example, as suspected or seen on the moons Enceladus, Europa, Titan, and Triton

cubewano Any large Kuiper belt object orbiting between about 41 AU and 48 AU but not controlled by orbital resonances with Neptune; the odd name is derived from 1992 QB_1, the first Kuiper belt object found

cyclone An area in the atmosphere in which the pressures are lower than those of the surrounding region at the same level, resulting in circular motion in a counterclockwise direction north of the equator and in a clockwise direction to the south

debris disk A flattened, spinning disk of dust and gas around a star formed from collisions among bodies already accreted in an aging solar system

differential rotation Rotation at different rates at different latitudes, requiring a liquid or gassy body, such as the Sun or Jupiter

differentiated body A spherical body that has a structure of concentric spherical layers, differing in terms of composition, heat, density, and/or motion; caused by gravitational separations and heating events such as planetary accretion

dipole Two associated magnetic poles, one positive and one negative, creating a magnetic field

direct (prograde) Rotation or orbit in the same direction as the Earth's, that is, counterclockwise when viewed from above its North Pole

disk wind Magnetic fields that either pull material into the protostar or push it into the outer disk; these are thought to form at the inner edge of the disk where the protostar's magnetic field crosses the disk's magnetic field (also called "x-wind")

distributary River channels that branch from the main river channel, carrying flow away from the central channel; usually form fans of channels at a river's delta

eccentricity The amount by which an ellipse differs from a circle

ecliptic The imaginary plane that contains the Earth's orbit and from which the planes of other planets' orbits deviate slightly (Pluto the most, by 17 degrees); the ecliptic makes an angle of 7 degrees with the plane of the Sun's equator

ejecta Material thrown out of the site of a crater by the force of the impactor

element A family of atoms that all have the same number of positively charged particles in their nuclei (the center of the atom)

ellipticity The amount by which a planet's shape deviates from a sphere

equinox One of two points in a planet's orbit when day and night have the same length; vernal equinox occurs in Earth's spring and autumnal equinox in the fall

exosphere The uppermost layer of a planet's atmosphere

extrasolar Outside this solar system

faint young Sun paradox The apparent contradiction between the observation that the Sun gave off far less heat in its early years, and the likelihood that the Earth was still warm enough to host liquid water

garnet The red, green, or purple mineral that contains the majority of the aluminum in the Earth's upper mantle; its compositional formula is $((Ca,Mg,Fe\ Mn)_3(Al,Fe,Cr,Ti)_2(SiO_4)3)$

giant molecular cloud An interstellar cloud of dust and gas that is the birthplace of clusters of new stars as it collapses through its own gravity

graben A low area longer than it is wide and bounded from adjoining higher areas by faults; caused by extension in the crust

granite An intrusive igneous rock with high silica content and some minerals containing water; in this solar system thought to be found only on Earth

half-life The time it takes for half a population of an unstable isotope to decay

hydrogen burning The most basic process of nuclear fusion in the cores of stars that produces helium and radiation from hydrogen

igneous rock Rock that was once hot enough to be completely molten

impactor A generic term for the object striking and creating a crater in another body

inclination As commonly used in planetary science, the angle between the plane of a planet's orbit and the plane of the ecliptic

iron meteorite Meteorites that consist largely of iron-nickel metal; thought to be parts of the cores of smashed planetesimals from early solar system accretion

isotope Atoms with the same number of protons (and are therefore the same type of element) but different numbers of neutrons; may be stable or radioactive and occur in different relative abundances

lander A spacecraft designed to land on another solar system object rather than flying by, orbiting, or entering the atmosphere and then burning up or crashing

lithosphere The uppermost layer of a terrestrial planet consisting of stiff material that moves as one unit if there are plate tectonic forces and does not convect internally but transfers heat from the planet's interior through conduction

magnetic moment The torque (turning force) exerted on a magnet when it is placed in a magnetic field

magnetopause The surface between the magnetosheath and the magnetosphere of a planet

magnetosheath The compressed, heated portion of the solar wind where it piles up against a planetary magnetic field

magnetosphere The volume of a planet's magnetic field, shaped by the internal planetary source of the magnetism and by interactions with the solar wind

magnitude See APPARENT MAGNITUDE

mantle The spherical shell of a terrestrial planet between crust and core; thought to consist mainly of silicate minerals

mass number The number of protons plus neutrons in an atom's nucleus

mesosphere The atmospheric layer between the stratosphere and the thermosphere

metal 1) Material with high electrical conductivity in which the atomic nuclei are surrounded by a cloud of electrons, that is, metallic bonds, or 2) In astronomy, any element heavier than helium

metallicity The fraction of all elements heavier than hydrogen and helium in a star or protoplanetary disk; higher metallicity is thought to encourage the formation of planets

metamorphic rock Rock that has been changed from its original state by heat or pressure but was never liquid

mid-ocean ridge The line of active volcanism in oceanic basins from which two oceanic plates are produced, one moving away from each side of the ridge; only exist on Earth

mineral A naturally occurring inorganic substance having an orderly internal structure (usually crystalline) and characteristic chemical composition

nucleus The center of the atom, consisting of protons (positively charged) and neutrons (no electric charge); tiny in volume but makes up almost all the mass of the atom

nutation The slow wobble of a planet's rotation axis along a line of longitude, causing changes in the planet's obliquity

obliquity The angle between a planet's equatorial plane to its orbit plane

occultation The movement of one celestial body in front of another from a particular point of view; most commonly the movement of a planet in front of a star from the point of view of an Earth viewer

olivine Also known as the gem peridot, the green mineral that makes up the majority of the upper mantle; its compositional formula is $((Mg, Fe)_2SiO_4)$

one-plate planet A planet with lithosphere that forms a continuous spherical shell around the whole planet, not breaking into plates or moving with tectonics; Mercury, Venus, and Mars are examples

opposition When the Earth is between the Sun and the planet of interest

orbital period The time required for an object to make a complete circuit along its orbit

pallasite A type of iron meteorite that also contains the silicate mineral olivine, and is thought to be part of the region between the mantle and core in a differentiated planetesimal that was shattered in the early years of the solar system

parent body The larger body that has been broken to produce smaller pieces; large bodies in the asteroid belt are thought to be the parent bodies of meteorites that fall to Earth today

perigee   As for perihelion but for any orbital system (not confined to the Sun)

perihelion   A distance; the closest approach to the Sun made in an object's orbit

planetary nebula   A shell of gas ejected from stars at the end of their lifetimes; unfortunately named in an era of primitive telescopes that could not discern the size and nature of these objects

planetesimal   The small, condensed bodies that formed early in the solar system and presumably accreted to make the planets; probably resembled comets or asteroids

plate tectonics   The movement of lithospheric plates relative to each other, only known on Earth

precession   The movement of a planet's axis of rotation that causes the axis to change its direction of tilt, much as the direction of the axis of a toy top rotates as it slows

primordial disk   Another name for a protoplanetary disk

prograde (direct)   Rotates or orbits in the same direction the Earth does, that is, counterclockwise when viewed from above its North Pole

proplyd   Abbreviation for a *protoplanetary disk*

protoplanetary disk   The flattened, spinning cloud of dust and gas surrounding a growing new star

protostar   The central mass of gas and dust in a newly forming solar system that will eventually begin thermonuclear fusion and become a star

radioactive   An atom prone to radiodecay

radio-decay   The conversion of an atom into a different atom or isotope through emission of energy or subatomic particles

red, reddened   A solar system body with a redder color in visible light, but more important, one that has increased albedo at low wavelengths (the "red" end of the spectrum)

reflectance spectra   The spectrum of radiation that bounces off a surface, for example, sunlight bouncing off the surface of as asteroid; the wavelengths with low intensities show the kinds of radiation absorbed rather than reflected by the surface, and indicate the composition of the surface materials

refractory   An element that requires unusually high temperatures in order to melt or evaporate; compare to volatile

relief (topographic relief)   The shapes of the surface of land; most especially the high parts such as hills or mountains

resonance   When the ratio of the orbital periods of two bodies is an integer; for example, if one moon orbits its planet once for every two times another moon orbits, the two are said to be in resonance

retrograde   Rotates or orbits in the opposite direction to Earth, that is, clockwise when viewed from above its North Pole

Roche limit   The radius around a given planet that a given satellite must be outside of in order to remain intact; within the Roche limit, the satellite's self-gravity will be overcome by gravitational tidal forces from the planet, and the satellite will be torn apart

rock   Material consisting of the aggregate of minerals

sedimentary rock   Rock made of mineral grains that were transported by water or air

seismic waves   Waves of energy propagating through a planet, caused by earthquakes or other impulsive forces, such as meteorite impacts and human-made explosions

semimajor axis   Half the widest diameter of an orbit

semiminor axis   Half the narrowest diameter of an orbit

silicate   A molecule, crystal, or compound made from the basic building block silica (SiO_2); the Earth's mantle is made of silicates, while its core is made of metals

spectrometer   An instrument that separates electromagnetic radiation, such as light, into wavelengths, creating a spectrum

stratosphere The layer of the atmosphere located between the troposphere and the mesosphere, characterized by a slight temperature increase and absence of clouds

subduction Movement of one lithospheric plate beneath another

subduction zone A compressive boundary between two lithospheric plates, where one plate (usually an oceanic plate) is sliding beneath the other and plunging at an angle into the mantle

synchronous orbit radius The orbital radius at which the satellite's orbital period is equal to the rotational period of the planet; contrast with synchronous rotation

synchronous rotation When the same face of a moon is always toward its planet, caused by the period of the moon's rotation about its axis being the same as the period of the moon's orbit around its planet; most moons rotate synchronously due to tidal locking

tacholine The region in the Sun where differential rotation gives way to solid-body rotation, creating a shear zone and perhaps the body's magnetic field as well; is at the depth of about one-third of the Sun's radius

terrestrial planet A planet similar to the Earth—rocky and metallic and in the inner solar system; includes Mercury, Venus, Earth, and Mars

thermosphere The atmospheric layer between the mesosphere and the exosphere

tidal locking The tidal (gravitational) pull between two closely orbiting bodies that causes the bodies to settle into stable orbits with the same faces toward each other at all times; this final stable state is called synchronous rotation

tomography The technique of creating images of the interior of the Earth using the slightly different speeds of earthquake waves that have traveled along different paths through the Earth

tropopause The point in the atmosphere of any planet where the temperature reaches a minimum; both above and below this height, temperatures rise

troposphere The lower regions of a planetary atmosphere, where convection keeps the gas mixed, and there is a steady decrease in temperature with height above the surface

viscosity A liquid's resistance to flowing; honey has higher viscosity than water

visual magnitude The brightness of a celestial body as seen from Earth categorized on a numerical scale; the brightest star has magnitude −1.4 and the faintest visible star has magnitude 6; a decrease of one unit represents an increase in brightness by a factor of 2.512; system begun by Ptolemy in the second century B.C.E.; see also apparent magnitude

volatile An element that moves into a liquid or gas state at relatively low temperatures; compare with refractory

x-wind Magnetic fields that either pull material into the protostar or push it into the outer disk; these are thought to form at the inner edge of the disk where the protostar's magnetic field crosses the disk's magnetic field (also called "disk wind")

Further Resources

Acuña, M. H., J. E. P. Connerney, et al. "Global Distribution of Crystal Magnetism Discovered by the Mars Global Surveyor MAG/ER Experiment." *Science* 284: 790–793, 1,999. One of the first and best scientific papers on the discovery of the magnetic field on Mars.

Andrews-Hanna, J. C., M. T. Zuber, and W. B. Banerdt. "The Borealis Basin and the Origin of the Martian Crustal Dichotomy." *Nature* 453: 1,212–1,216. The newest version of the giant impact origin for the Martian northern lowlands: A single glancing blow.

Barlow, N. *Mars: An Introduction to Its Interior, Surface, and Atmosphere*. Cambridge: Cambridge University Press, 2008. Scientifically up-to-date summary of the history of Mars science and its current state, with numerous color images.

Beatty, J. K., C. C. Petersen, and A. Chaikin. *The New Solar System*. Cambridge: Sky Publishing and Cambridge University Press, 1999. The most commonly used textbook on the solar system, with comprehensive and scholarly articles written in accessible language.

Bell, J. *Postcards from Mars: The First Photographer on the Red Planet*. New York: Penguin Books, 2006. Collection of color photos taken by the Mars rovers and the story line of their creation by the head photographer on the rover team.

Booth, N. *Exploring the Solar System*. Cambridge: Cambridge University Press, 1995. Well-written and accurate volume on solar system exploration.

Comins, Neil F., and William J. Kaufmann. *Discovering the Universe*. New York: W. H. Freeman, 2008. The best-selling text for astronomy courses that does not use mathematics. Presents concepts clearly and stresses the process of science.

Dickin, A. P. *Radiogenic Isotope Geology*. Cambridge: Cambridge University Press, 1995. Clear and complete textbook on the study of isotopes in geological sciences.

Ferris, J. C., J. M. Dohm, V. R. Baker, and T. Maddick III. "Dark Slope Streaks on Mars: Are Aqueous Processes Involved?" *Geophysical Research Letters* 29 (2002): 128. An early paper on the discovery of possible groundwater leakage on Mars.

Fradin, Dennis Brindell. *The Planet Hunters: The Search for Other Worlds*. New York: Simon and Schuster, 1997. Stories of the people who through time have hunted for and found the planets.

Frey, H. V., J. H. Roark, K. M. Shockey, E. L. Frey, and S. E. H. Sakimoto. "Ancient Lowlands on Mars." *Geophysical Research Letters* 29 (2002): 22. Describes the original theory that the northern lowlands of Mars might be created by several large impact craters.

Jakosky, Bruce M., and Michael T. Mellon "Water on Mars." *Physics Today* (April 2004): 71–76. Description of the state of knowledge of water on Mars, written for a lay scientific audience.

Kleine, T., C. Munker, K. Mezger, and H. Palme. "Rapid Accretion and Early Core Formation on Asteroids and the Terrestrial Planets from Hf-W Chronometry." *Nature* 418: 952–955. The science of dating the time of core formation, early in solar system history.

Norton, R. O. *The Cambridge Encyclopedia of Meteorites*. Cambridge: Cambridge University Press, 2002. Beautifully illustrated and up-to-date book on meteorites.

Paul, N. *The Solar System*. Secaucus, N.J.: Chartwell Books, 2008. Begins with the origin of the universe and moves through the planets. Includes history of space flight and many color images.

Rees, Martin. *Universe*. London: DK Adult. A team of science writers and astronomers wrote this text for high school students and the general public.

Sleep, N. H. "Martian Plate Tectonics." *Journal of Geophysical Research* 90: 5,639–5,655. Was there once plate tectonics on Mars? A story of the evidence and the science.

Sparrow, Giles. *The Planets: A Journey through the Solar System*. Waltham, Mass.: Quercus Press, 2009. Solar system discoveries told within the structure of the last 40 years of space missions.

Spence, P. *The Universe Revealed*. Cambridge: Cambridge University Press, 1998. Comprehensive textbook on the universe.

Stacey, Frank D. *Physics of the Earth*. Brisbane, Aust.: Brookfield Press, 1992. Fundamental geophysics text on an upper-level undergraduate college level.

Stevenson, D. J. "Planetary magnetic fields." *Earth and Planetary Science Letters* 208: 1–11 (2003). Comparisons and calculations about the planetary magnetic fields of many bodies in our solar system.

Zuber, M. T. "The Crust and Mantle of Mars." *Nature* 412: 220–227. A summary of the knowledge of the outer shells of Mars, written for a lay scientific audience.

INTERNET RESOURCES

Arnett, Bill. "The ~~Nine~~ 8 Planets: A Multimedia Tour of the Solar System." Available online. URL: http://nineplanets.org. First accessed September 21, 2009. An accessible overview of the history and science of the eight planets and their moons.

Blue, Jennifer, and the Working Group for Planetary System Nomenclature. "Gazetteer of Planetary Nomenclature." United States Geological Survey. Available online. URL: http://planetarynames.wr.usgs.gov/. Accessed May 27, 2009. Complete and official rules for naming planetary features, along with list of all named planetary features and downloadable images.

European Space Agency. "Lessons Online: Human Spaceflight and Exploration." Available online. URL: http://www.esa.int/esaMI/Lessons_online/SEMUY1V7D7F_0.html. Accessed May 26, 2008. The space radiation environment and its affects on humans.

Goodall, Kirk, and the Mars Global Surveyor Science Team. "Mars Global Surveyor." Jet Propulsion Laboratory. Available online. URL: http://mars.jpl.nasa.gov/mgs/index.html. Accessed May 29, 2009. Official home page of the Mars Global Surveyor mission, including press releases, mission description, and images.

Harvey, Ralph P. "ANSMET: The Antarctic Search for Meteorites." Available online. URL: http://geology.cwru.edu/~ansmet/. Accessed December 19, 2009. Explains the history, motivation, and logistics of the annual expedition to search for meteorites on the Antarctic ice sheet. Also carries blogs from the current expedition.

LaVoie, Sue, Myche McAuley, and Elizabeth Duxbury Rye. "Planetary Photojournal," Jet Propulsion Laboratory and NASA. Available online. URL: http://photojournal.jpl.nasa.gov/index.html. Accessed May 27, 2009. Large database of public-domain images from space missions.

O'Connor, John J., and Edmund F. Robertson. "The MacTutor History of Mathematics Archive." University of St. Andrews, Scotland. Available online. URL: http://www-gap.dcs.st-and.ac.uk/~history/index.html. Accessed May 27, 2009. A scholarly, precise, and eminently accessible compilation of biographies and accomplishments of mathematicians and scientists through the ages.

Rowlett, Russ. "How Many? A Dictionary of Units of Measurement." University of North Carolina at Chapel Hill. Available online. URL: http://www.unc.edu/~rowlett/units. Accessed May 27, 2009. A comprehensive dictionary of units of measurement, from the metric and English systems to the most obscure usages.

Smith, David E. "Mars Orbital Laser Altimeter (MOLA) Science Investigation," NASA. Available online. URL: http://ssed.gsfc.nasa.gov/tharsis/mola.html. Accessed May 27, 2009. Complete science results from the MOLA instrument on the Mars Global Surveyor mission.

Smith, Peter H. "Phoenix Mars Mission." NASA and Arizona. Available online. URL: http://phoenix.lpl.arizona.edu/timeline.php. Accessed May 27, 2009. Official home page for the *Phoenix* mission. Includes a beautifully laid-out page with records and links to all the Mars missions.

United States Navy. "U.S. Naval Observatory Astronomical Applications Department." Available online. URL: http://aa.usno.navy.mil/. Accessed September 21, 2009. Astronomical data for practical applications; includes lists of equinoxes, almanacs, and software.

Viotti, Michelle. "Mars Exploration Rover Mission." NASA and JPL. Available online. URL: http://marsrovers.jpl.nasa.gov/home/. Accessed May 27, 2009. Official and complete home page for the Mars Exploration rovers, including press releases, science information, and images.

————. "Mars Reconnaissance Orbiter." NASA and JPL Available online. URL: http://marsprogram.jpl.nasa.gov/mro/. Accessed May 27, 2009. Home page for the Mars Reconnaissance Orbiter mission.

White, Maura, and Allan Stilwell. "JSC Digital Image Collection." Johnson Space Center. Available online. URL: http://www.nasa.gov/multimedia/imagegallery/. Accessed May 27, 2009. A catalog of over 9,000 NASA press release photos from the entirety of the manned space flight program.

Williams, David. "Planetary Fact Sheets." NASA. Available online. URL: http://nssdc.gsfc.nasa.gov/planetary/planetfact.html. Accessed September 21, 2009. Detailed measurements and data on the planets, asteroids, and comets in simple tables.

Williams, David, and Dr. Ed Grayzeck. "Lunar and Planetary Science." NASA. Available online. URL: http://nssdc.gsfc.nasa.gov/planetary/planetary_home.html. First accessed September 21, 2009. NASA's deep archive and general distribution center for lunar and planetary data and images.

ORGANIZATIONS OF INTEREST

American Geophysical Union (AGU)

2000 Florida Avenue N.W.

Washington, DC 20009-1277

USA

www.agu.org

AGU is a worldwide scientific community that advances, through unselfish cooperation in research, the understanding of Earth and space for the benefit of humanity. AGU is an individual membership society open to those professionally engaged in or associated with the Earth and space sciences. Membership has increased steadily each year, doubling during the 1980s. Membership currently exceeds 41,000, of which about 20 percent are students. Membership in AGU entitles members and associates to receive Eos, *AGU's weekly newspaper, and* Physics Today, *a magazine produced by the American Institute of Physics. In addition they are entitled to special member rates for AGU publications and meetings.*

Association of Space Explorers

1150 Gemini Avenue

Houston TX 77058

http://www.space-explorers.org/

This association is expressly for people who have flown in space. They include 320 individuals from 34 nations, and their goal is to support space science and education. Their outreach activities include a speakers program, astronaut school visits, and observer status with the United Nations.

European Space Agency (ESA)

8–10 rue Mario Nikis

75738 Paris, Cedex 15

France

http://www.esa.int/esaCP/index.html

The European Space Agency has 18 member states, which together create a unified European space program and carry out missions in parallel and in cooperation with NASA, JAXA, and other space agencies. Its member countries are Austria, Belgium, Czech Republic, Denmark, Finland, France, Germany, Greece, Ireland, Italy, Luxembourg, the Netherlands, Norway, Portugal, Spain, Sweden, Switzerland, and the United Kingdom. Hungary, Romania, Poland, and Slovenia are cooperating partners.

International Astronomical Union (IAU)

98 bis, Boulevarde Arago, 75014

Paris, France

www.iau.org

The International Astronomical Union (IAU) was founded in 1919. Its mission is to promote and safeguard the science of astronomy in all its aspects through international cooperation. Its individual members are professional astronomers all over the world, at the Ph.D. level or beyond and active in professional research and education in astronomy. However, the IAU also maintains friendly relations with organizations that include amateur astronomers in their membership. National members are generally those with a significant level of professional astronomy. With now over 9,100 individual members and 65 national members worldwide, the IAU plays a pivotal role in promoting and coordinating worldwide cooperation in astronomy. The IAU also serves as the internationally recognized authority for assigning designations to celestial bodies and any surface features on them.

Jet Propulsion Laboratory (JPL)

4800 Oak Grove Drive

Pasadena, CA, 91109

www.jpl.nasa.gov

The Jet Propulsion Laboratory is managed by the California Institute of Technology for NASA. JPL manages many of NASA's space missions, including the Mars Rovers and Cassini and also conducts fundamental research in planetary and space science.

Meteoritical Society: The International Society for Meteoritics and Planetary Science

www.meteoriticalsociety.org

The Meteoritical Society is a nonprofit scholarly organization founded in 1933 to promote the study of extraterrestrial materials and their history. The membership of the society includes 950 scientists and amateur enthusiasts from over 33 countries who are interested in a wide range of planetary science. Members' interests include meteorites, cosmic dust,

*asteroids and comets, natural satellites, planets, impacts, and the ori-
gins of the solar system.*

National Aeronautics and Space Administration (NASA)
300 E Street S.W.
Washington DC 20002
www.nasa.gov
*NASA, an agency of the U.S. government, manages space flight centers,
research centers, and other organizations including the National Aerospace Museum. NASA scientists and engineers conduct basic research
on planetary and space topics, plan and execute space missions, oversee
Earth satellites and data collection, and many other space- and flight-
related projects.*

Planetary Society
65 North Catalina Avenue
Pasadena CA 91106-2301
http://www.planetary.org/home/
*This is a society of lay individuals, scientists, organizations, and busi-
nesses dedicated to involving the world's public in space exploration
through advocacy, projects, and exploration. The Planetary Society was
founded in 1980 by Carl Sagan, Bruce Murray, and Louis Friedman.
They are particularly dedicated to searching for life outside of Earth.*

Index